Site Carpentry

Peter Brett

D1581211

Nelson Thornes

First edition published in 1993 by: Stanley Thornes (Publishers) Ltd.

Second edition published in 2002

This edition published in 2007 by:
Nelson Thornes Ltd
Delta Place
27 Bath Road
CHELTENHAM
GL53 7TH
United Kingdom

10 11 12 13 / 10 9 8 7 6 5 4 3 2

A catalogue record for this book is available from the British Library

ISBN 978 0 7487 8185 0

Cover photos by : Brand X HC 274 (NT), Corbis CI 183 (NT), Ingram ILS V1 CD2 CON (NT), Ingram ILR V1 CD2 BS (NT), Pixland/Jupiter 178 (NT)

New illustrations by Peters and Zabransky Ltd
Achive illustrations include artwork by Peters and Zabransky (UK) Ltd and Richard Morris, with colour added by GreenGate Publishing Services.

Page make-up by GreenGate Publishing Services, Tonbridge, Kent

Printed in China by 1010 Printing International Ltd

Contents

Acknowledgements

My sincere thanks go to: my wife Christine for her assistance, support and constant encouragement; my daughter Sarah; grandchildren Matthew, Chris and Rebecca and son James and his partner Claire for their support, patience and sanity checks; my colleagues and associates past and present for their continued support of my work and motivation to continue.

Finally 'all the best for the future' to all who use this book, I trust it provides you with some of the help and motivation required to succeed in the construction industry.

Peter Brett

Photograph credits:

Brand X HC 274 (NT), pp.73, 104

Corbis CI 183 (NT), Measuring up, Activity, Did you know?, Safety tip icons, p.164

Elizabeth Whiting Associates, p.175

George Disario/Corbis, p.ix

Ingram ILS V1 CD2 CON (NT), p.215

Martin Sookias, with thanks to Oldham College for use of their wood machine workshop, p.217

Pixland/Jupiter 178 (NT), pp.1, 8

Rentokil Property Care and David Cropp, pp.173, 174

Warwickshire College, with thanks, for the use of their furniture making workshop, pp.218, 219

Introduction

National Vocational Qualifications (NVQs) in Construction

These qualifications focus on practical skills and knowledge. They have been developed and approved by people that work in the construction industry.

Construction NVQs are available in England, Wales and Northern Ireland. Scotland uses SVQs, which work in a similar way.

There are three levels of NVQs for construction crafts and operatives:

◆ Level 1 is seen as a 'foundation' to the construction industry, consisting of common core skills and occupational basic skills.
◆ Level 2 consists of common core skills and units of competence in a recognizable work role.
◆ Level 3 consists of further common core skills, plus a more complex set of units of competence in a recognizable work role, including some work of a supervisory nature.

Awarding body

CITB-Construction Skills and City & Guilds are the joint awarding body for the construction industry. CITB-Construction Skills are also responsible for the setting of standards for Craft and Operative NVQs.

Work roles

Each construction NVQ focuses on an individual work role. For example:

◆ bricklaying;
◆ site carpentry;
◆ bench joinery;
◆ painting and decorating;
◆ plastering;
◆ shop fitting etc.

Construction NVQ make up

Each work role is made up of a number of individual units of competence. For example:

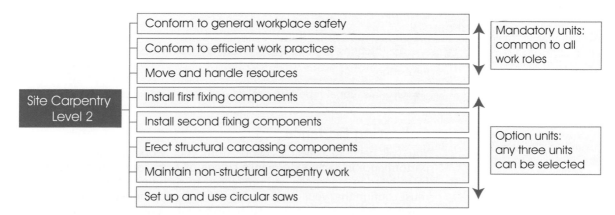

Figure 0.1 *Qualification structure*

All mandatory units must be undertaken plus a number of the options (three in the case of site carpentry). The number of option units in a work role and the number that are required to be undertaken will vary depending on the extent of the particular work role.

Unit of competence make up: in order to set out exactly what is contained in a unit and also make it easier to assess, each unit begins with a description, for example:

Conform to General Workplace Safety

This unit is about:

◆ awareness of relevant current statutory requirements and official guidance;
◆ personal responsibilities relating to workplace safety, wearing appropriate personal protective equipment (PPE) and compliance with warning/safety signs;
◆ personal behaviour in the workplace;
◆ security in the workplace.

The description is followed by a number of statements:

Performance criteria: these state exactly what you must be able to do.

Identify hazards

Scope of performance: this sets out what evidence is required to meet each of the performance criteria. The majority of this evidence must be from the workplace, simulation evidence is only allowed in limited circumstances.

Hazards, associated with the workplace and occupations at work, are recorded and/or reported

Knowledge and understanding relating to performance criteria: this links in general terms the knowledge and understanding required to back up the performance criteria.

You must know and understand:

◆ the hazards associated with the occupational area;
◆ the method of reporting hazards in the workplace.

Scope of knowledge and understanding: this uses the key words contained in the Knowledge and Understanding statements (shown in bold type) and expands them to cover the scope of what is expected of a competent worker in the construction industry.

Hazards:

◆ Associated with resources, workplace, environment, substances, equipment, obstructions, storage, services and work activities.

Reporting:

◆ Organisational reporting procedures and statutory requirements.

Collecting Evidence

You will need to collect evidence from your workplace of your satisfactory performance in each performance criteria of a unit of competence. This should be inserted in a portfolio and referenced to each unit of competence. Evidence must confirm that your practical skills meet the appropriate performance criteria. Simulation evidence in a training environment is only allowed in a limited range of topics.

Evidence can come from any of the following people:

◆ employers;
◆ managers;
◆ supervisors;
◆ skilled work colleagues;
◆ work-based recorder;
◆ client.

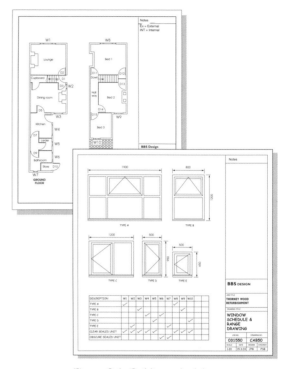

Figure 0.2 *Evidence of 'me' wearing PPE*

Figure 0.3 *Extracts from a specification*

Figure 0.4 *Building schedule*

Suitable types of evidence – you should include in your portfolio as much evidence as possible, from more than one of the following, for each performance criteria.

◆ Time sheets: detailing the work you have undertaken, for these to be valid they must be signed by yourself and the work-based recorder.
◆ Drawings: of the work you have undertaken. These should be supported by a witness testimony.
◆ Photographs: of the work you have undertaken, ideally with you in the photograph. To be valid photographs should be supported by a statement, containing a brief description of the work, details of where and when you carried it out and be signed by yourself and either the work-based recorder, manager, supervisor, skilled worker or the client.
◆ Associated documentation: used or produced as part of the work you have undertaken, such as specifications, forms and reports completed.
◆ Witness testimony: a statement by a responsible person confirming that you have undertaken certain work activities, these should include wherever possible a detailed description of the work you carried out.

Introduction

Introduction

T. Joycee Construction

Ridge House
Norton road
Cheltenham
GL 59 1DB

To whom it may concern:

I can confirm, that between 15 March and 8 November 2006 James Oakley worked on the refurbishment contract at The Rivermead Estate.

James was involved in the replacement of casement windows and internal window boards. He carried this work out to a competent standard at all times.

This work was undertaken in occupied houses, feedback from the tenants concerning James's communication with them and his consideration shown to their property, including the cleanliness of his work was always exemplary.

In addition James assisted me in the general day-to-day organization of the working environment, including the scheduling of the work and the safety induction of new staff. Indeed he always set a fine example by wearing at all times his safety helmet, boots and high visibility vest.

Yours Faithfully

Chris Heath

Chris Heath
(Site Project Manager)

Figure 0.5 *Witness testimony*

Where an assessor considers your evidence as insufficient in either quality or quantity, you may be asked to undertake simulated activities in order to demonstrate/reinforce your competence in particular performance criteria.

The assessment process

The joint awarding body CITB-Construction Skills and City & Guilds approves organisations to carry out assessment of people for an NVQ award in construction. Typically these are:

◆ further education colleges;
◆ private training providers;
◆ construction companies.

Once approved these are known as assessment organizations. Their assessment work will involve the following personnel:

◆ *Assessors* – these are people who are occupationally competent in the work role in which you are being assessed and also qualified in the assessment process. Their role is to decide whether you are competent in each performance criterion. They will also observe you in the workplace to ensure you are carrying out the full range of activities to create required evidence portfolio.
◆ *Internal verifier* – this is the person who is responsible, in an assessment organisation, for ensuring the quality of the assessments carried out by the assessors.
◆ *Work-based recorders* – these are people in the workplace whose employer has given them the responsibility of authenticating the evidence that a candidate is collecting for a portfolio.
◆ *External verifiers* – they are employed by the joint awarding body to monitor the whole assessment process and ensure that each assessment organization is working to the standards set.

How to use this book

This book covers the five occupational-specific skills units for site carpentry at Level 2. Separate books are available for the wood occupations at Level 1 and site carpentry at Level 2. The mandatory common core units are covered in a companion book, *A Building Craft Foundation*, to which reference should be made.

These books are intended to be supported by:

◆ classroom activities;
◆ tutor reinforcement and guidance;
◆ group discussion;
◆ films, slides and videos;
◆ text books;
◆ independent study/research;
◆ practical activities.

You will be working towards one or more units at a time as required. Discuss each unit's content with your group, tutor, or friends wherever possible. Attempt to answer the learning activities for that unit. Progressively work through all the units, discussing them and answering the assessment activities as you go. At the same time you should be working on the matching practical activities in the workplace and collecting the required evidence.

This process is intended to aid learning and enable you to evaluate your understanding of the particular topic and to check your progress through the units. Where you are unable to answer a question, further reading and discussion of the topic are required.

Independent study/research

'Browsing the Internet' via a computer is an excellent means of accessing other sources of information as part of your research: simply type in the website address of the company or organisation into a web browser and you will be connected to their website.

Try some of the following sites:

• Building Regulations: www.planningportal.gov.uk;
• British Standards: www.bsi-global.com;
• Building Research Establishment: ww.bre.co.uk;
• construction training and careers: www.citb-constructionskills.co.uk and www.city-guilds.co.uk;
• Government publications: www.tso.co.uk;
• health and safety: www.hse.gov.uk;
• building materials and components: www.buildingcentre.co.uk;
• types and use of timber in construction: www.trada.co.uk;
• employment rights and trade unions www.worksmart.org.uk.

If you don't know the exact website address of the organisation you are looking for, or you simply wish to find out more information on a subject, you could use a 'search engine' to find the web pages. Search engines use 'key words' to find information on a subject. Enter a key word or words such as doors, windows, stairs or strength grading or timber etc. or the name of a company/organization, and it searches the Internet for information about your key words or name. You are then presented with a list of relevant websites that you can click on, which link you to the appropriate information pages.

Types of learning activity

The learning activities used in this book should be completed on loose-leaf paper and included as part of your portfolio of evidence. They are divided into the following:

◆ Measuring up. Questions at the end of a major topic or units, which enable you to evaluate your understanding of a recently completed topic and to check your progress through the units. 'Measuring-up' questions are either multiple-choice questions or short answer questions.
◆ Activity. An extended learning task normally at the end of a unit, which has been designed to reinforce your technical and communication skills in day-to-day work situations.

Multiple-choice questions normally consist of a statement or question followed by four possible answers. Only one answer is correct; the others are distracters. You have to select the most appropriate letter as your response.

Example 1:

The joint use between a rafter and wallplate in a pitched roof is termed a:

(a) butt
(b) birdsmouth
(c) dovetail
(d) seat

As birdsmouth is the correct answer your response should be (b).

Example 2:

The flight of stairs illustrated is called:

(a) open
(b) closed
(c) freestanding
(d) alternating

As 'closed' is the correct answer, your response should be (b).

Occasionally variants on the four-option multiple-choice question are used, as in the following examples.

Example 3:

Match the items in *list one* with the items in *list two*.

List one:

1. Pencil rounded
2. Splayed and rounded
3. Ovolo
4. Ogee
5. Combination profile
6. Torus

List two: refer to illustration

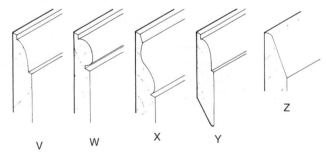

V   W   X   Y   Z

Choose the correct match:

	V	W	X	Y	Z
(a)	3	6	4	5	1
(b)	4	3	6	2	1
(c)	4	6	5	1	2
(d)	3	6	4	5	2

This question requires you to work through the lists matching the items up (it is usual for the lists to be of different lengths). In this example:

V is 3
W is 6
X is 4
Y is 5
Z is 2

Therefore the correct response is (d).

Example 4:

Statement: pipe casings in kitchens and bathrooms are best made using a WBP plywood.

Reason: WBP means that the plywood has been manufactured using a weather and boil proof adhesive.

(a) statement true	reason true
(b) statement false	reason false
(c) statement true	reason false
(d) statement false	reason true

This type of question consists of a statement followed by a reason, where the both the statement and reason can be true or false. You are required to select the appropriate response.

In this example both the statement and reason are true, therefore the correct response is (a).

Short-answer questions consist of a task to which a short written answer is required. The length will vary depending on the 'doing' word in the task:

◆ 'name' or 'list' normally require one or two words for each item;
◆ 'state', 'define', 'describe' or 'explain' will require a short sentence or two;
◆ 'draw' or 'sketch' will require you to produce an illustration.

In addition, sketches can be added to any written answer to aid clarification.

Example 5:

Name the documents that give details of door type and ironmongery required for a particular situation.

Typical answer: door and ironmongery schedules.

Example 6:

State the reason why a pre-made studwork partition is made smaller than the space in which it is to be fitted.

Typical answer: to allow for a positioning tolerance.

Example 7:

Produce a sketch to show the difference between herringbone and solid strutting to floor joists.

Typical answer:

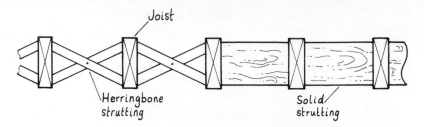

Figure 0.6 *Difference between herring-bone and solid strutting to floor joists*

'Activity' sections are a combination of short-answer questions on the same topic. They normally commence with a statement containing a certain amount of background information designed to set the scene for the question. This is then followed by a series of questions in logical order. The length of the expected answer to each sub-part will vary, depending on the topic and the wording of the question, from one or two words to a paragraph. There may be a blank form to complete, a sketch, a calculation, or a combination of any of these. At each stage the wording of the question will make it clear what is required. Blank forms for completion and inclusion in your portfolio of evidence may be downloaded from www. Nelsonhomes.com/carpentry.

Example 8:

The illustration below shows details of a timber-suspended floor:

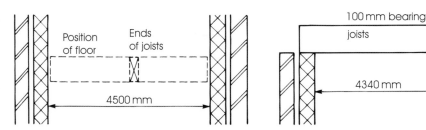

(a) Name the joist that spans from wall to wall.
(b) What do the Building Regulations require to be fixed at mid-span to prevent any buckling or sagging in large span joists?
(c) What is the effective span of the joists?
(d) Determine the number of 50 mm breath joists required to be spaced at approximately 400 mm centres, between the two walls.
(e) Determine the total length of timber required in metres run for all the floor joists.

Typical answer:

(a) A joist that spans from wall to wall is known as a bridging joist.
(b) The Building Regulations require strutting to be fixed at the mid-span to prevent buckling or sagging in large span joists.
(c) The effective span of a joist is the distance between the centres of the joist bearings; in this case 4440 mm.
(d) It is standard practice to position the outer joists 50 mm away from the walls. The centres of the outer 50 mm breadth joists would be 75 mm away from the walls. Thus the distance between the outer joists centres is 4350 mm.
Number of joists $= (4350 \div 400) + 1$

$$= 10.875 + 1$$

$$= 11.875$$

Thus 12 joists are required.

(e) Total length of timber required $= 12 \times 4540$

$$= 54.480\,m$$

Thus total length required is 54.480 m run.

In addition to completing the learning activities, you may be asked oral questions by your tutor, assessor or verifier. This is often done to gain further evidence of your written response or questions may be asked during a review of your portfolio to gain supplementary evidence: these questions normally take the form of 'How did you…?' 'Why did you…?' 'What would you do in the following circumstance…s?' etc.

Other learning features used in this book

These include the following:

Colour enhanced illustrations and documents as an aid to clarity and reinforcement of text.

Did you know boxes in the margin, which define new words or highlight key facts.

did you know?

The use of machined timber is recommended for carcassing as it has a consistent-sized section.

Safety Tip boxes in the margin, highlighting facts for you to follow or be aware of when undertaking practical tasks.

Example

Mortise deadlock, latch or lock/latch

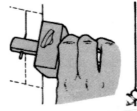

Mark position on door edge

Gauge centre line on door edge

Drill out to width and depth

Worked examples included in the text for use as a guide when answering questions or undertaking other tasks.

safety tip

Always ensure that the machine is isolated from the power supply before making any adjustments to guards, before changing any tooling, and before undertaking maintenance or cleaning down.

example

A 3.114 m length of wall is to be panelled with 95 mm (90 mm covering width) matchboard. Divide the length of wall in millimetres by the covering width of the board.

$$3114 \div 90 = 34.6 \text{ boards}$$

Therefore 35 boards are required = 33 whole boards and two end boards.

Width of cut end boards $= 1.6 \times 90 \div 2$

$$= 72\,mm$$

Introduction

Underpinning Skills

This chapter is intended to provide the entrant at Level 2 with a review of the enabling skills and supporting job knowledge required successfully to complete the main practical activities in each of the Site Carpentry Level 2 units. Although its content is not assessed directly, knowledge of its content is assumed and assessed in other units. It is concerned with the range of underpinning skills that are encountered on a day-to-day basis.

In this chapter you will cover the following range of topics:

◆ interpreting instructions and planning own work;
◆ adopting safe working practices;
◆ identifying, maintaining and using hand tools;
◆ setting up and using portable power tools;
◆ handling timber-based materials and components;
◆ recognition and use of timber and associated products;
◆ calculations.

You may have already achieved some or all of these skills and knowledge either in industry or as a result of training at Level 1 or similar. Thus this underpinning skills chapter has been included in the form of typical questions for you to answer. Questions are divided into topic areas. Where you cannot answer any particular question, further study should be undertaken using either the information source indicated, other appropriate textbooks, or talk it through with your tutor or a workmate.

This chapter can be studied on its own or alongside other chapters according to your need.

Persons with prior achievement may wish to use these questions on underpinning skills as a refresher to support other chapters as required.

Interpreting instructions and planning own work

These two topics are covered in *A Building Craft Foundation* (3rd edition) under 'Communications' and 'Materials'. These should be referred to if you have difficulty in answering the following questions.

measuring up

1. State the reason why construction drawings are drawn to a scale and not full size.

2. Mark on the scale rule shown below 4.550 m to a scale of 1:50.

[scale rule diagram: JAKAR, Metric, 315 PL., British Made; markings 1:5 0, 100mm, 200, 300, 400, 500, 600, 700mm; 1:50 1m, 2, 3, 4, 5, 6, 7m; lower scales 1:2500 and 1:1250]

3. Produce sketches to show the standard symbols used to represent: brickwork, blockwork, concrete, sawn and planed timber.

4. State what is meant by orthographic projection.

5. Define the terms 'plan', 'elevation' and 'section' when applied to a drawing of an object.

6. State the purpose of specifications and schedules.

7. State why messages must be relayed accurately.

8. State the meaning of the following standard abbreviations:

   bwk   bldg
   DPC   dwg
   hwd   swd

9. State the action to be taken when damaged goods are received from a supplier.

10. State ONE reason why you as an employee should plan how to carry out work given to you.

11. Name the person you should contact in the event of a technical problem occurring at work.

12. State the reason why dust sheets should be used when working internally in occupied premises.

13. State why it is important to be polite with the customer.

14. State why it is important to be co-operative and helpful with work colleagues.

did you know?

The terms 'unwrot' and 'wrot' are sometimes used instead of sawn and planed timber.

Adopting safe working practices

This topic is covered in *A Building Craft Foundation* (3rd edition) under 'Health and Safety'. This should be referred to if you have difficulty in answering the following questions.

15. State TWO duties expected of you as an employee under the Health and Safety at Work Act.

16. State TWO objectives of the Health and Safety at Work Act.

17. State TWO main powers of a Health and Safety Executive Inspector.

18. State TWO situations where protective equipment must be used. Name the item of equipment in EACH case.

19. State the reason for keeping work areas clear and tidy.

20. Name a suitable fire extinguisher for use on a flammable liquid or gas fire.

21. Describe the correct body position for lifting a large box from ground level.

22. Name the type of safety sign that is contained in a yellow triangle with a black border.

23. Describe the role of a site Safety Officer.

24. State the purpose of a toe board on a scaffold platform.

25. Define the terms 'hazard' and 'accident'.

26. List THREE checks that should be made before using a scaffold.

27. State the correct working angle of a ladder.

28. Briefly explain the procedures to be followed in the event of an emergency incident occurring on site.

29. The greatest number of fatal accidents in the construction industry involve:
    (a)   Falls from a height
    (b)   Being struck by moving object
    (c)   Machinery
    (d)   Electricity.

Identifying, maintaining and using hand tools

This topic is covered in *Wood Occupations* (2nd edition). This should be referred to if you have difficulty in answering the following questions.

30. Produce a sketch to show the difference in cutting action between a rip and cross-cut saw.

31. Name the saw best used for cutting architrave mitres.

32. Define the difference between a warrington and claw hammer.

33. State an advantage of using a water level over using a spirit level.

34. State the procedure used for sharpening a plane iron.

35. When sharpening saws the following operations are carried out: setting, shaping, sharpening and topping. State the order in which these are carried out.

36. Explain the operations carried out when preparing a piece of sawn timber to PAR by hand.

37. State the purpose of using oil when sharpening plane and chisel blades.

38. Name the type of work for which a panel saw is most suitable.

39. Name the type of work for which a bullnose plane is most suitable.

40. Name THREE different types of chisel and state a use for EACH.

41. State the purpose of a bradawl.

42. Produce a sketch to show a mitre template and state a situation where it may be used.

43. Name a tool that can be used to draw large diameter curves.

44. State the reason for taking off the corners of a smoothing plane iron after sharpening.

Setting up and using portable power tools

This topic is covered in *Wood Occupations* (2nd edition). This should be referred to if you have difficulty in answering the following questions.

measuring up

45. State FOUR basic safety rules that should be followed when using any powered tool.

46. When using a hand-held electric circular saw state THREE operations that should be carried out before plugging the tool into the power supply.

47. State the reason why power tools should never be carried, dragged or suspended by their cables.

48. Describe the procedure for plunge cutting with a jig saw.

49. State the reason why the cutters of a portable planer should be allowed to stop before putting the tool down.

50. State how cutters are held in a portable powered router.

51. Describe the THREE basic work stages when using a plunging portable powered router.

52. Produce a sketch to show the correct direction of feed for a router in relation to the rotation of the cutter.

53. State the purpose of using 110 volt power tools.

54. What type of power tool does not require an earth wire?

55. State the correct location of an extension cable used in conjunction with a transformer that steps 240 volts mains supply down to 110 volts.

56. Which of the following sanders is best used for fine finishing work: circular, orbital, belt?

57. A cartridge-operated fixing tool is to be used for fixing timber grounds to a concrete ceiling. List THREE items of equipment that are recommended for the operator to wear.

58. The cartridge-operated fixing tool you are using on site has been supplied with THREE colours of cartridge: red, black and yellow. List them in order of decreasing strength.

59. State the action that the operator should take if a power tool is not working correctly or its safety is suspect.

Handling timber-based materials and components

This topic is covered in *A Building Craft Foundation* (3rd edition) under 'Materials'. This should be referred to if you have difficulty in answering the following questions.

60. State the reason for stacking timber off the ground.

61. State TWO reasons why materials storage on site should be planned.

62. State THREE personal hygiene precautions that may be recommended by a manufacturer when handling materials.

63. State the reason for using piling sticks or cross-bearers when stacking carcassing timber.

64. Give the reason for stacking sheet materials flat and level.

65. Explain why joinery should be stored under cover after delivery.

66. State the reason why the leaning of items of joinery against walls is not to be recommended.

67. State the reason why new deliveries are put at the back of existing stock in the store.

68. Explain why liquids should not be kept in any container other than that supplied by the manufacturer.

69. Explain why veneered sheets of plywood are stored good face to good face.

Recognition and use of timber and associated products

◆ Timber and manufactured boards.
◆ Preservatives.
◆ Adhesives.
◆ Fixings.

This topic is covered in *Wood Occupations* (2nd edition). This should be referred to if you have difficulty in answering the following questions.

Underpinning Skills

Chapter 1

70. Describe THREE main differences between softwoods and hardwoods.

71. List FOUR common sawn sizes for carcassing timber.

72. Softwood is available in stock lengths from 1.8 m. State the measurement that stock lengths increase by.

73. Produce a sketch to distinguish between multi-ply, blockboard and laminboard.

74. Describe what is meant by stress-graded timber and name TWO grades.

75. Produce sketches to show the following mouldings: torus, ogee, bullnosed, ovolo, scotia.

76. List the TWO initial factors that must be present for an attack of dry rot in timber.

77. Describe the THREE stages of an attack of dry rot.

78. State the purpose of using preservative-treated timber.

79. Name TWO types of timber preservative and state TWO methods of application.

80. Produce sketches to show the following timber defects: cup shake, knot, cupping, waney edge and sloping grain.

81. Name TWO common wood-boring insects and for EACH state the location and timber they will most likely attack.

82. Define what is meant by conversion of timber and produce sketches to show through-and-through and quarter sawn.

83. Define the term 'seasoning' of timber.

84. State a suitable moisture content when installing carcassing timber and explain how a moisture meter measures this.

85. State TWO advantages that sheet material has over the use of solid timber.

86. A sheet of plywood has been marked up WBP grade. Explain what this means.

87. Explain the reason why water is brushed into the mesh side of hardboard prior to its use.

88. Define the following terms when applied to adhesives: storage/shelf life; pot life.

89. Explain the essential safety precaution to be taken when using a contact adhesive.

90. Produce a sketch to show the difference between countersunk, round-head and raised-head screws.

91. Describe a situation where EACH of the following nails may be used: wire nail, oval nail, annular nail and masonry nail.

92. Define with the aid of sketches EACH of the following types of nailing: dovetail, skew and secret.

93. Describe a situation where a non-ferrous metal plug would be specified for screwing into rather than a fibre or plastic plug.

Calculations

This topic is covered in *A Building Craft Foundation* (3rd edition) under 'Numerical Skills'. This should be referred to if you have difficulty in answering the following questions.

94. Add together the following dimensions 750 mm, 1.200 m, 705 mm and 4.645 mm.

95. 756 joinery components are produced by a manufacturer: 327 are to be preservative treated, the remainder require painting. State how many are to be painted.

96. Nine pieces of timber are required to make an item of joinery. How many pieces of timber are required to make 17 such items?

97. A rectangular room measures 4.8 m × 5.2 m. Calculate the floor area and the length of skirting required. Allow for ONE 900 mm wide door opening.

98. Five semi-circular pieces of plywood are required. Calculate the cost of plywood at £4.55 per square metre if the radius of EACH semi-circular piece is 600 mm.

99. A 105-m run of carcassing timber is required for a project. You have been asked to allow an additional 15% for cutting and wastage. Determine the amount to be ordered.

100. A triangular piece of plywood has a base span of 1.4 m and a rise of 500 mm. Determine in square metres the area of five such pieces.

101. A semi-circular bay window has a diameter of 2.4 m. Determine the length of skirting required for this window.

102. A door 1980 mm in height is to have a handle fixed centrally. A security viewer is to be positioned 350 mm above this height. Determine the height of the viewer.

103. A carpenter earns £98.40 per day. The apprentice is paid 30% of this amount. Determine the wage bill for five days for both people if the employer has to allow an additional 17.5% for on-costs.

104. Use a calculator or tables to solve the following:
    (a)   457 divided by 239
    (b)   6945 multiplied by 1350
    (c)   336 raised to the third power
    (d)   the square root of 183

Carcassing

This chapter is intended to provide the reader with an overview of carcassing work. Its contents are assessed in NVQ Unit No. VR 11 Erect Structural Carcassing Components.

In this chapter you will cover the following range of topics:

◆ floor and flat roof joists;
◆ joist coverings;
◆ roof types;
◆ cut roof erection;
◆ verge and eaves finishings.

What's required in VR 11 Erect Structural Carcassing Components?

To successfully complete this unit you will be required to demonstrate your skill and knowledge of the following carcassing elements:

◆ floors, roofs and their finishings.

You will be required practically to:

◆ construct inclined roofs with gables;
◆ install roof verge and eaves finishings;
◆ position joists for ground and upper floors or flat roofs;
◆ lay flat roof decking or flooring.

Note: The laying of flat roof decking or flooring is also a requirement of NVQ Unit No. VR 09 Install First Fixing Components. Once suitable evidence has been obtained it can be used for both units.

did you know?

Carcassing is the process of building the carcass of a building, which includes the walls, floors and roofs.

Floor and flat roof joists

Joist terminology

Floor

The floor is the ground or upper levels in a building, which provides an acceptable surface for walking, living and working.

Timber ground floor – the floor of a building nearest the exterior ground level and known as a hollow or suspended floor. Joists are supported at intervals by honeycomb sleeper walls. Air bricks and ventilation gaps in the sleeper walls provide ventilation to the underfloor space to keep the timber dry and reduce the possibility of rot.

Timber upper floors – the floor levels of a building above the ground floor. They are known as suspended floors. Bridging joists span between supports. Binders may be incorporated to reduce span; strutting is used in mid-span to reduce tendency to buckle. Openings in floors are framed using trimming, trimmer and trimmed joists.

Roof

The roof is the uppermost part of a building that spans the external walls and provides protection from the elements.

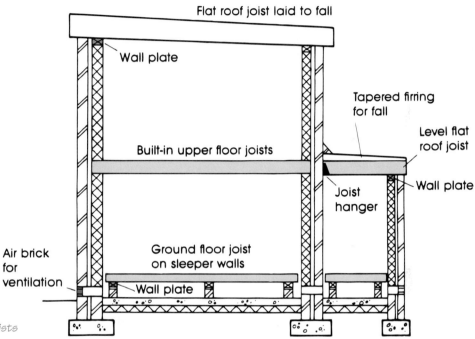

Figure 2.1 *Floor and roof joists*

Timber flat roofs – any roof having an angle or slope that is less than 10 degrees to the horizontal. They are constructed similar to timber upper floors. The slope on the top surface may be formed by either laying the bridging joist to falls (out of level), or by the use of firrings.

Joist

A joist is one of a series of parallel timber beams, used to span the gap between walls and directly support a floor surface, ceiling surface or flat roof surface.

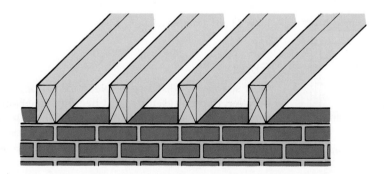

Figure 2.2 *Joists*

The sectional size of a joist depends on its span, spacing, weight or loading placed upon it and the quality of the timber used. The Timber Research and Development Association (TRADA) and other bodies produce tables of suitable sectional joist sizes and spacing for use in various situations. See Table 2.1 for a typical example.

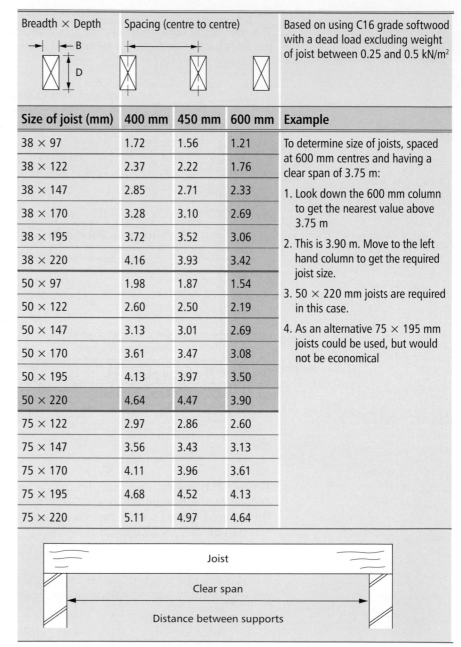

Breadth × Depth	Spacing (centre to centre)			Based on using C16 grade softwood with a dead load excluding weight of joist between 0.25 and 0.5 kN/m²
Size of joist (mm)	400 mm	450 mm	600 mm	Example
38 × 97	1.72	1.56	1.21	To determine size of joists, spaced at 600 mm centres and having a clear span of 3.75 m:
38 × 122	2.37	2.22	1.76	
38 × 147	2.85	2.71	2.33	1. Look down the 600 mm column to get the nearest value above 3.75 m
38 × 170	3.28	3.10	2.69	
38 × 195	3.72	3.52	3.06	2. This is 3.90 m. Move to the left hand column to get the required joist size.
38 × 220	4.16	3.93	3.42	
50 × 97	1.98	1.87	1.54	3. 50 × 220 mm joists are required in this case.
50 × 122	2.60	2.50	2.19	
50 × 147	3.13	3.01	2.69	4. As an alternative 75 × 195 mm joists could be used, but would not be economical
50 × 170	3.61	3.47	3.08	
50 × 195	4.13	3.97	3.50	
50 × 220	4.64	4.47	3.90	
75 × 122	2.97	2.86	2.60	
75 × 147	3.56	3.43	3.13	
75 × 170	4.11	3.96	3.61	
75 × 195	4.68	4.52	4.13	
75 × 220	5.11	4.97	4.64	

Table 2.1 *Maximum clear span of floor joists (m)*

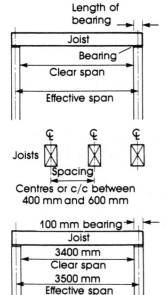

Figure 2.3 *Joist span and spacing*

Span

Clear span is the distance between joist supports. Effective span is the distance between the centres of the joist bearings (see Figure 2.3). The bearing itself is the length of the end of a joist that rests on the support. The overall length of a joist is its clear span plus the length of its end bearings, e.g. a joist with a 3400 mm clear span and 100 mm end bearings will have an effective span of 3500 mm and an overall joist length of 3600 mm.

Joists are commonly laid out to span the shortest distance between the supporting walls of a room or other area. This keeps to a minimum the size of joist required. Once the depth of the joist is determined for the longest span, it is normal practice to keep all other joists the same. The shorter span joists will be oversize, but all joist covering and ceiling surfaces will be level.

Spacing

Joist spacing is the distance between the centres of adjacent joists. Commonly called joist centres or c/c (centre to centre), they range between 400 to 600 mm depending on the joist covering material.

Joists should be spaced to accommodate surface dimensions of their covering material. Table 2.2 gives details of the maximum span for a range of joist covering materials. These maximum spans will dictate the maximum joist spacing, depending on the joist covering material used. End joists adjacent to walls should be kept 50 mm away from the wall surface, in order to allow an air circulation and prevent dampness being transferred from wall to joist, and thus reduce the risk of rot. In addition this gap helps reduce the transmission of noise at party walls.

Decking material	Finished thickness	Maximum span
		Span = joist spacing
Softwood planed, tongued and grooved P, T & G	16 mm	450 mm
Softwood planed, tongued and grooved P, T & G	19 mm	600 mm
Flooring grade chipboard	18 mm	450 mm
Flooring grade chipboard	22 mm	600 mm
Flooring grade plywood	16 mm	400 mm
Flooring grade plywood	19 mm	600 mm

Table 2.2 *Guide to maximum span of decking (joist covering materials)*

Determining the number of joists

To determine the number of joists required and their centres for a particular area the following procedure, shown in Figure 2.4, can be used:

◆ Measure the distance between adjacent walls, say 3150 mm.
◆ The first and last joist would be positioned 50 mm away from the walls. The centres of 50 mm breadth joists would be 75 mm away from the wall. The total distance between end joists centres would be 3000 mm.
◆ Divide distance between end joists centres by specified joist spacing say 400 mm. This gives the number of spaces between joists. Where a whole number is not achieved, round up to the nearest whole number above. There will always be one more joist than the number of spaces, so add one to this figure to determine the number of joists.
◆ Where tongue-and-groove boarding is used as a floor covering the joist centres may be spaced out evenly, i.e. divide the distance between end joist centres by the number of spaces.

Carcassing **Chapter 2**

◆ Where sheet material is used as a joist covering to form a floor, ceiling or roof surface, the joist centres are normally maintained at a 400 mm or 600 mm module to coincide with sheet sizes. This would leave an undersized spacing between the last two joists.

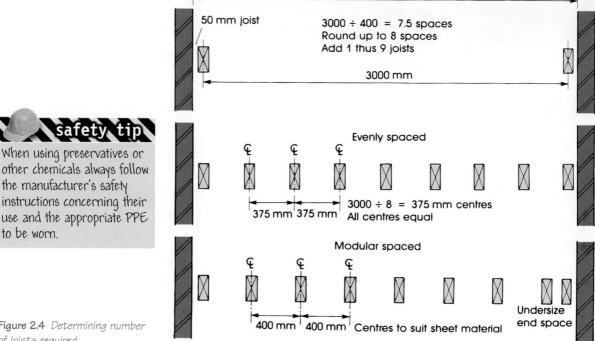

3150 mm

50 mm joist

3000 ÷ 400 = 7.5 spaces
Round up to 8 spaces
Add 1 thus 9 joists

3000 mm

Evenly spaced

3000 ÷ 8 = 375 mm centres
375 mm 375 mm All centres equal

Modular spaced

400 mm 400 mm Centres to suit sheet material

Undersize end space

Figure 2.4 *Determining number of joists required*

Preservative treatment

It is recommended that all timber used for structural purposes is preservative treated before use. Any preservative-treated timber cut to size on site will require re-treatment on the freshly cut edges/ends. This can be carried out by applying two brush flood coats of preservative. Timber preservatives prevent rot by poisoning the food supply on which fungi feed and grow.

activity

Determine the number of 50 mm breadth joists required to be spaced at approximately 400 mm centres, between two walls 4350 mm apart.

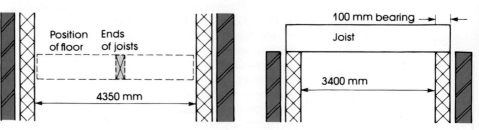

Position of floor   Ends of joists

4350 mm

100 mm bearing

Joist

3400 mm

Figure 2.5 *Upper floor sections*

Determine the total amount of timber required in metres run, for all the floor joists. These have a clear span of 3400 mm and a bearing at each end of 100 mm.

measuring up

1. State the purpose of a joist.

2. Why do joists normally span the shortest distance?

3. Define a joist's clear span.

4. The total length of joist having an effective span of 3600 mm and end bearings of 100 mm is:
   (a) 3500 mm
   (b) 3600 mm
   (c) 3700 mm
   (d) 3800 mm

5. If a joist spans between two walls 2.8 m apart and has end bearings of 100 mm, what length of joist is required?

6. Name the regulations that apply to the positioning and fixing of joists.

7. State the reason for treating sawn ends of joists with preservative.

Fixing joists

Section

A section is the breadth and depth of a joist. The strength of a joist varies in direct proportion to changes in its breadth and in proportion to the square of its depth. For example, doubling the breadth of a section doubles its strength.

A 100 mm × 100 mm joist has double the strength of a 50 mm × 100 mm joist, whereas doubling the depth of a section increases its strength by four times.

A 50 mm × 200 mm joist has four times the strength of a 50 mm × 100 mm joist.

Less material is required for the same strength when the greatest sectional dimension is placed vertically rather than horizontally.

A 100 mm × 100 mm joist has the same sectional area as a 50 mm × 200 mm joist, but the deeper joist would be twice as strong.

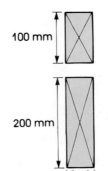

100 mm

200 mm

50 mm

Doubling depth
increases strength
by four times

$100^2 = 100 \times 100 = 10\,000$
$200^2 = 200 \times 200 = 40\,000$

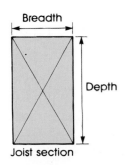

Breadth

Depth

Joist section

Figure 2.6 *Joist section*

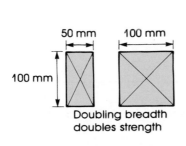

50 mm   100 mm

100 mm

Doubling breadth
doubles strength

Carcassing **Chapter 2**

Therefore joists are normally placed so that the depth is the greatest sectional dimension. Joists of the same sectional size and span would clearly have different strengths if their breadths and depths were reversed. Those with the smaller depth would sag under load, possibly leading to structural collapse (see Figure 2.7). When stating the sectional size of a joist the first measurement given is the breadth and the second the depth.

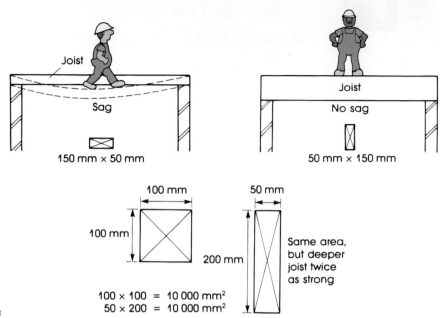

Figure 2.7 *Positioning of joist section*

Machined, formerly termed 'regularized', timber, as shown in Figure 2.8, is preferred for joists as all timber will be a consistent depth. This aids levelling and ensures a flat fixing surface for joist coverings and ceilings. Joists are machined to a consistent depth by re-sawing or planing one or both edges. A reduction in size of 3 mm must be allowed for timber up to 150 mm in depth and 5 mm, over this width.

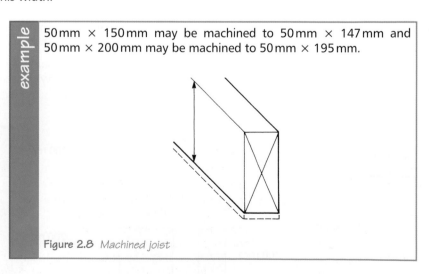

example

50 mm × 150 mm may be machined to 50 mm × 147 mm and 50 mm × 200 mm may be machined to 50 mm × 195 mm.

Figure 2.8 *Machined joist*

Joist supports

Any joists that are not straight should be positioned with their camber or crown (curved edge) upwards. When loaded these joists will tend to straighten out rather than sag further if laid the other way. Joists with edge knots should be positioned with the knots on their upper edge. When loaded the knots, if on their upper edge, will be held in position as the joists sag, rather than fall out, weaken the joist and possibly lead to structural collapse if laid the other way (see Figure 2.9).

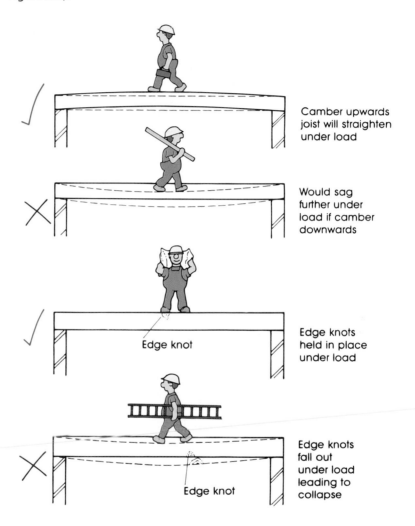

Camber upwards joist will straighten under load

Would sag further under load if camber downwards

Edge knots held in place under load

Edge knots fall out under load leading to collapse

Edge knot

Edge knot

Figure 2.9 Positioning of joists with camber or edge knots

The ends of joists may be supported:

◆ by building in;
◆ on hangers;
◆ on wall plates;
◆ on binders.

Building in – ends of joints are supported on the inner leaf of a cavity wall (see Figure 2.10). The minimum bearing in a wall is normally 90 mm. (Shorter bearings do not tie in the wall sufficiently and can lead to a crushing of the joist end possibly leading to collapse.) A steel bearing bar may be incorporated into the mortar joint where lightweight blocks are used to reduce the risk of the blocks crumbling. The ends of the joists are often splayed but they must not project into the cavity where they could possibly catch mortar droppings during building, leading to dampness in the joist and rotting. The ends of joists that are in contact with the external wall should be treated with a timber preservative to protect them from dampness and subsequent rot.

> **did you know?**
>
> Joist hangers made from thin metal, used for timber-to-timber joints, do not normally require recessing in.

TIMBER FLOOR

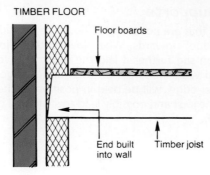

Figure 2.10 *Building in a joist*

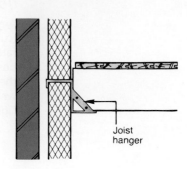

Figure 2.11 *Use of joist hanger*

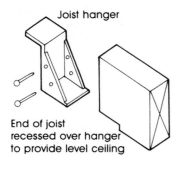

Figure 2.12 *Recessing end of joist*

Hangers – ends of joists may be supported on galvanized steel joist hangers, which are built into or bear on a wall (see Figure 2.11). Double hangers are available that saddle internal walls to provide a support for joists on both sides. An advantage of this method is that the joist can be positioned independently of the building process. Hangers are useful when forming extensions as they are simply inserted into a raked out mortar joint. The bottom edge or bearing surface of the hanger must be recessed into the joist, as shown in Figure 2.12. This ensures that the top and bottom edges of the joists are flush and also prevent hangers obstructing any ceiling covering. Joists should be secured into the hanger using 32 mm galvanized clout nails in each hole provided.

Wall plates – ends of joists may be supported by a wall plate bedded on the top of a wall. This is normally used for ground floor construction, flat roofs and internal load-bearing partitions. The minimum bearing required and thus the minimum width of wall plate is 75 mm. Often joists from either side meet over a wall plate; it is usual to nail them together side by side, both overlapping the wall plate by about 300 mm.

The use of wall plates provides a means of securing joists by skew nailing with 75 mm or 100 mm wire nails. In addition, wall plates also spread the loading of a joisted surface evenly over a wide area rather than a point load. Cambered joists may be straightened over a wall plate by partly sawing through and nailing down. Wall plates are not suitable for use in external walls of upper floor construction. This is due to shrinkage movement and the likelihood of rot.

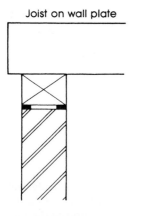

Figure 2.13 *Joist supported on a wall plate*

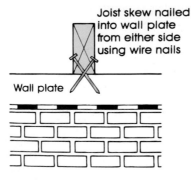

Figure 2.14 *Fixing joist to wall plate*

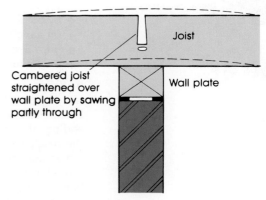

Figure 2.15 *Straightening cambered joist over wall plate*

Wall plates are jointed by the carpenter using halving joints as shown in Figure 2.16. The plates are bedded and levelled in position by the bricklayer using bedding mortar.

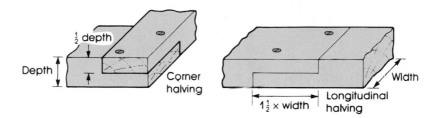

Figure 2.16 *Wall plate joints*

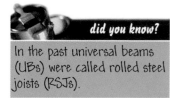

did you know?

Floors with binders are known as double floors.

Binders – binders are introduced into a floor structure in order to provide an intermediate support for large span joists used in double and triple floors. These binders may be of:

◆ steel, when they are known as a universal beam (UB);
◆ timber, either of solid section, glue laminated section (glulam) or a plywood box beam; or
◆ concrete.

Depending on the space available, binders may be positioned below the joists, or accommodated partly within the joist depth projecting above or below as required. See Figure 2.17.

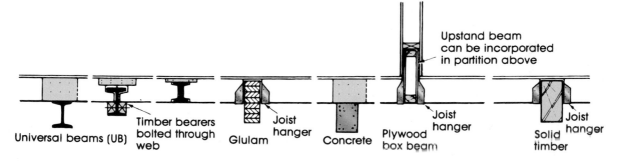

Figure 2.17 *Binders*

Where joists are fitted to a steel universal beam, a plywood template may be cut to speed the marking out of the joist end and ensure a consistent, accurate fit. (See Figure 2.18.)

did you know?

In the past universal beams (UBs) were called rolled steel joists (RSJs).

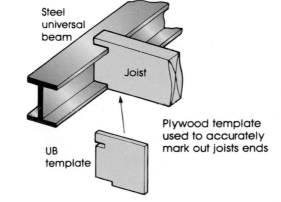

Figure 2.18 *Marking and cutting a joist to fit a universal beam*

Fire protection of steel binders – where universal beams are used in floor construction they must be protected as a fire precaution. This protection is required in order to prevent early structural collapse resulting from the fact that the strength of steel rapidly decreases at temperatures above 300°C. Various methods of protection include:

◆ a sprayed coat of non-combustible insulating material;
◆ a solid casing of concrete;
◆ encasing in expanded metal lathing and plastered on;
◆ encasing with plasterboard or other non-combustible material.

Figure 2.19 shows a typical method of encasing a steel beam using plasterboard and a skim coat of finishing plaster. The plasterboard has been fixed to timber cradling members nailed to each joist.

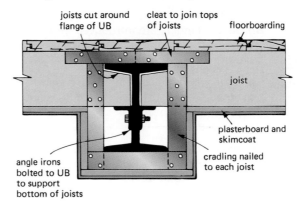

Figure 2.19 *Plasterboard and cradling to a universal beam*

Joint restraint – restraint straps

With modern lightweight structures, walls and joisted areas require positive tying together for strength, to ensure wall stability in windy conditions. Galvanized mild steel restraint straps, also known as lateral restraint straps, must be used at not more than 2 m intervals and carried over at least three joists. Noggins should be used under the straps where the joists run parallel to the wall; these should be a minimum of 38 mm thick and extend at least half the depth of the joist. For example, with 200 mm deep joists at least 100 mm deep noggins are required. See Figure 2.20.

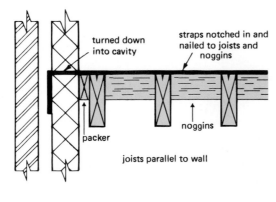

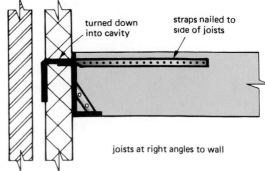

Figure 2.20 *Restraint strap*

Joint restraint – strutting

Where deep joists exceed a span of 2.5 m, they tend to buckle and/or sag under loads unless restrained by strutting. Joists spanning between 2.5 and 4.5 m should be restrained by strutting at their mid span; larger span joists over 4.5 m will require two rows of strutting at one-third and two-thirds of their span. There are three main types of strutting in use (see Figure 2.21):

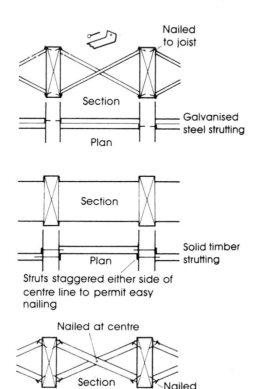

Figure 2.21 *Strutting to joists*

◆ Solid strutting is quick to install but is considered inferior, as it tends to loosen and become ineffective when joists shrink. It should be at least 38 mm thick and extend to at least three-quarters of the joist's depth, e.g. a 200 mm deep joist will require at least 150 mm-deep struts.

◆ Herringbone timber strutting is considered the most effective as it actually tightens when joists shrink. However, it takes longer to install and thus is more expensive in terms of labour costs. The minimum sectional size is 38 mm × 38 mm. Timber herringbone struts can only be used where the spacing between the joists is less than three times the depth of the joist, e.g. for 150 mm deep joists herringbone struts are suitable up to 450 mm spacing, but for 200 mm deep joists a spacing of up to 600 mm is permitted.

◆ Galvanized steel herringbone strutting is quick to install as no cutting is required, but has a disadvantage in that the depth and centres of the joist must be specified when ordering; different depths and spacing will require different sized struts.

Herringbone timber struts can be marked out using the following procedure shown in Figure 2.22.

1. Mark across the joists the centre line of the strutting.
2. Mark a second line across the joist so that the distance between the lines is 10 mm less than the depth of the joist.

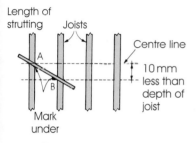

Figure 2.22 *Marking out strutting*

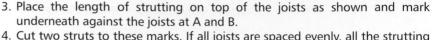

3. Place the length of strutting on top of the joists as shown and mark underneath against the joists at A and B.
4. Cut two struts to these marks. If all joists are spaced evenly, all the strutting will be the same size and can be cut using the first one as a template. If not, each set of struts will have to be marked individually.
5. Fix struts on either side of the centre line using wire nails, one in the top and bottom of each strut and one through the centre.

Whichever method of strutting is used, care should be taken to ensure that they are clear of the tops and bottoms of the joists; otherwise they may subsequently distort the joist covering or ceiling surface.

did you know?

Folding wedges and end packing pieces should be treated with preservative.

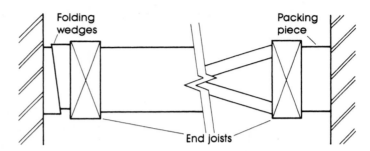

Figure 2.23 *Strutting tightened using a packing piece and folding wedges*

Again, whichever method of strutting is used, the gaps between the end joists and the walls will require packing and wedging to complete the system as illustrated in Figure 2.23. Care must be taken not to over tighten the folding wedges, as it is possible to dislodge the blockwork.

measuring up

8. Two pieces of timber are of the same sectional area, one is 100 mm × 100 mm and the other is 50 mm × 200 mm.
  (a) Which is best for use as a joist?
  (b) State why.

9. Use a sketch to define regularized timber.

10. State why joists should not project into wall cavities.

11. State the minimum joist bearing when:
  (a) building in
  (b) on wall plates.

12. State why regularized joists are preferred.

13. State a reason for recessing joists to receive joist hangers.

14. State the total metres run of timber required for six joists each spanning 3.6 m and being supported by joist hangers.

15. State the purpose of binders.

16. State the purpose of restraint straps.

17. At what centres should restraint straps be fixed?

18. State the purpose of a template when cutting joists into a steel binder.

19. Produce sketches to illustrate the following:
    - (a) a joist hanger
    - (b) a binder
    - (c) a wall plate lengthening (longitudinal) joint
    - (d) a restraint strap.

20. State the purpose of strutting joists.

21. What is the disadvantage of solid strutting?

22. Name the type of nail used to fix timber herringbone strutting.

23. State why strutting should be kept clear of the joist's top and bottom edges.

Trimming

Where openings are required in joisted areas or where projections occur in supporting walls, joists must be framed or trimmed around them (see Figure 2.24). Members used for trimming each have their own function and are named accordingly.

Bridging joist – a joist spanning from support to support, also known as a common joist.

Trimmed joist – a bridging joist that has been cut short (trimmed) to form an opening in the floor.

Trimmer joist – a joist placed at right angles to the bridging joist, in order to support the cut ends of the trimmed joists.

Trimming joist – a joist with a span the same as the bridging joist, but supporting the end of a trimmer joist.

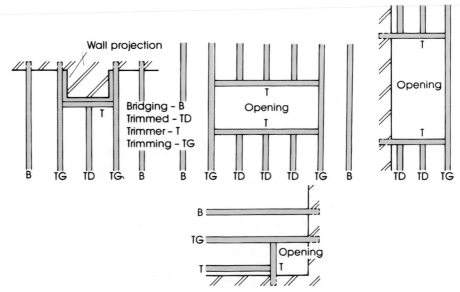

Figure 2.24 *Trimming openings*

Carcassing

Chapter 2

As both the trimmer and the trimming joists take a greater load, they are usually 25 mm thicker in breadth than the bridging joists. For example, use 75 mm × 200 mm trimmer and trimming joists with 50 mm × 200 mm bridging joists.

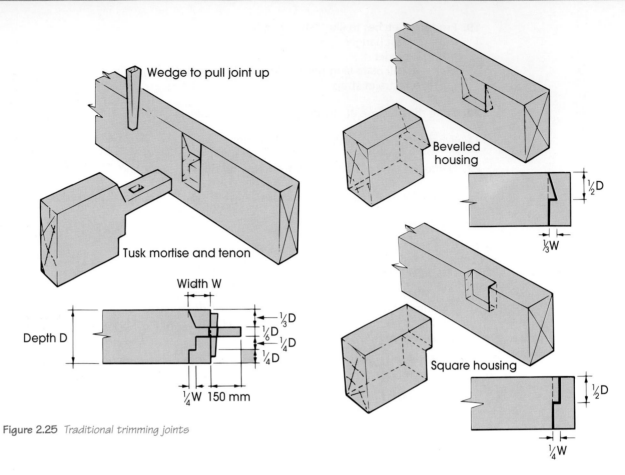

Wedge to pull joint up

Bevelled housing

½D

⅓W

Tusk mortise and tenon

Width W

Depth D

⅓D
⅙D
¼D
¼D

¼W 150 mm

Square housing

½D

¼W

Figure 2.25 *Traditional trimming joints*

Trimming joints

Traditionally, tusk mortise and tenon joints were used between the trimmer and trimming joists, while housing joints were used between the trimmed joists and the trimmer as illustrated in Figure 2.25. Once wedged the tusk mortise and tenon joint requires no further fixing. The housing joints will require securing with 100 mm wire nails. The proportions of these joints must be followed. They are based on the fact that there are neutral stress areas in joists (see Figure 2.26).

Joints cut in shaded areas will cause the minimum reduction in strength

Figure 2.26 *Neutral stress areas*

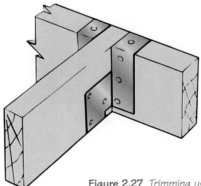

Figure 2.27 *Trimming using a joist hanger*

If any cutting is restricted mainly to this area then the reduction in the joist strength will be kept to a minimum. Joist hangers, which are a quicker, modern alternative to the traditional joints, are now used almost exclusively. As these hangers are made from thin galvanized steel, they do not require recessing in as do the thicker wall hangers but they must be securely nailed in each hole provided with 32 mm galvanized clout nails (see Figure 2.27).

Positioning joists

Vertical position of joists

The bricklayers will normally have established this. They will have built up to the required brickwork course for the underside of built-in floor joists or bedded a wall plate on at this position. When joist hangers are being used the brickwork is built up to the required course to take the lug of the hanger. This will be near the top edge of the joist. See Figure 2.28. Alternatively where straight lugged hangers, rather than the turned down restraint hanger, are used, the wall may be built beyond the floor level and the bed joint is raked out to receive the hangers.

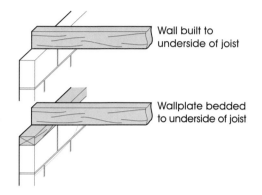

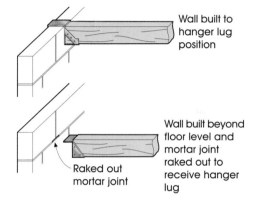

Figure 2.28
Vertical position of joists

Laying out joists

Joists are first cut to length (see Figure 2.29). Ground floor joists should be cut between the walls to give a gap of about 20 mm at both ends. Upper floor joists that are built in should be cut with a splayed end to avoid any possibility of the end protruding into the cavity. Joists in hangers must be cut to the exact length between hangers to keep the hangers tight against the wall. Take care not to cut them too long, as forcing them in can easily push the freshly laid walls out of position.

The outside joists are placed in position first, leaving a 50 mm gap between them and the wall. The other joists are then spaced out to the required centres in the remaining area.

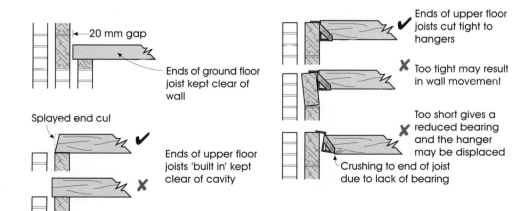

Figure 2.29 *Cutting joists to length*

However, when openings for fireplaces and stairwells etc. are required in the joisted area the layout of the joists is governed by these openings. The trimming and outside bridging joists are the first to be positioned, once again leaving a 50 mm gap between all joists and the walls. The other joists are then spaced out to the required centres in the remaining areas as illustrated in Figure 2.30.

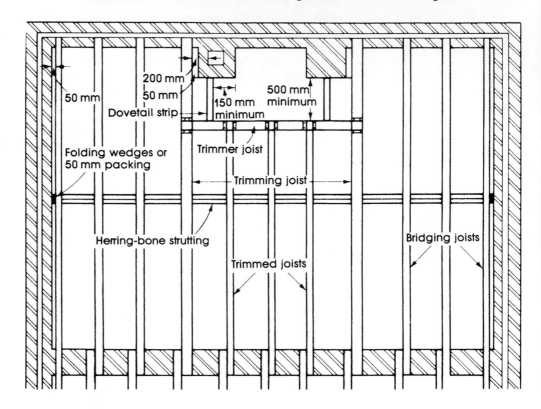

Figure 2.30
Layout of floor around fireplace

Timber near sources of heat

The Building Regulations restrict the positioning and trimming of timber near sources of heat. They state that the construction around heat-producing appliances should minimize the risk of the building catching fire. In order to comply with this requirement the use of combustible material in the vicinity of a fireplace, chimney or flue is restricted (see Figure 2.31).

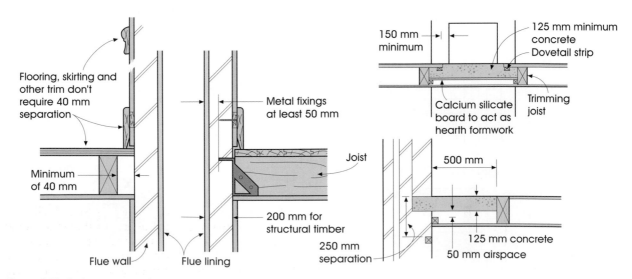

Figure 2.31 *Timber construction near a source of heat*

◆ No timber is to be built into the flue, or be within 200 mm of the flue lining, or be nearer than 40 mm to the outer surface of a chimney or fireplace recess. (A 50 mm airspace between any joist and wall is standard practice.) The exceptions to this requirement are: flooring; skirting; architrave; mantelshelf; and other trim. However, any metal fixings associated with these must be at least 50 mm from the flue.

◆ There must be a 125 mm minimum thickness solid non-combustible hearth, which extends at least 500 mm in front of the fireplace recess and 150 mm on both sides.

◆ Combustible material used underneath the hearth is to be separated from the hearth by airspace of at least 50 mm or be at least 250 mm from the hearth's top surface. Combustible material supporting the edge of the hearth is permitted.

Figure 2.32 illustrates a method of trimming upper floor joists around a projecting flue, from a ground floor heat-producing appliance.

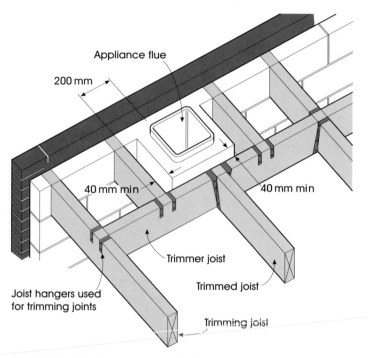

Figure 2.32 *Trimming around a projecting flue*

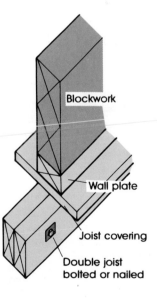

Figure 2.33 *Double joists required to support blockwork partition above*

Double joists

Double joists are required where blockwork partition walls are to be built on a joisted area. These double joists are either nailed or bolted together (depending on the specification) and positioned under the intended wall to take the additional load as illustrated in Figure 2.33. When laying out a joisted area, double joists are positioned along with the trimming and end bridging joists, before the other joists are spaced.

Change of joist direction

In order to span the shortest direction it is sometimes necessary to change the direction of joists in a particular joisted area. This is done over a load-bearing wall where the bridging joists from one area are allowed to overhang by 50 mm; the end-bridging joist in the other area is simply nailed to the overhanging ends (see Figure 2.34).

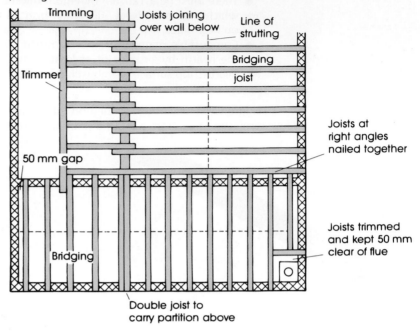

Figure 2.34 *Typical joist layout*

Levelling joists

After positioning, the joists should be checked for line and level. End joists are set using a spirit level; intermediate joists are lined through with a straight edge and spirit level (see Figure 2.35).

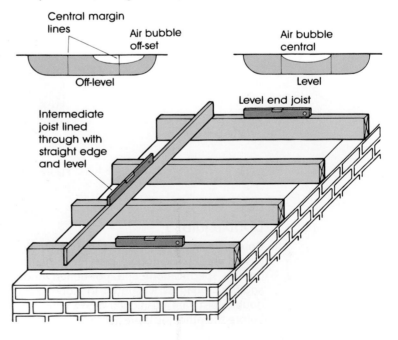

Figure 2.35 *Levelling joists*

Where the brickwork or blockwork course has been finished to a level line by the bricklayer or where wall plates are used and have been accurately bedded level, and where machined joists have been used, the bearings of the joist should not require any adjustment to bring them into line and level. However, minor adjustments may be required as shown in Figure 2.36. Joists may be housed into or packed off wall plates. Where packings are required for built-in joists, these should be slate or other durable material. Do not use timber packings as these may shrink and work loose and in any case are susceptible to rot. Joists may be recessed, to lower their bearing, providing the reduced joist depth is still sufficient for its span.

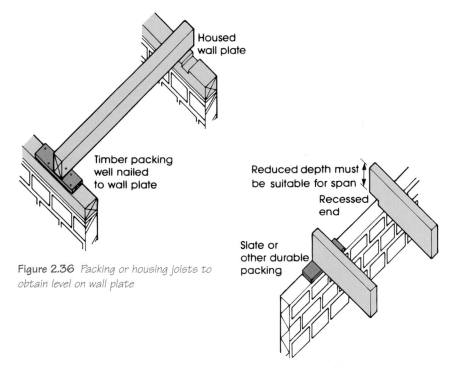

Figure 2.36 *Packing or housing joists to obtain level on wall plate*

Figure 2.37 *Packing or recessing joists to obtain level when built-in*

Figure 2.38 shows temporary battens that can be nailed across the top of the joists to ensure that their spacing remains constant before and during their 'building in'. Joists fixed to wall plates can be skew nailed to them using 75 mm or 100 mm wire nails.

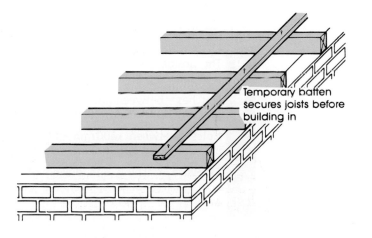

Figure 2.38 *Use of temporary batten to secure joists before building in*

did you know?

Notches and holes in joists to accommodate services should be kept to a minimum to avoid excessive reduction in strength.

Notching joists

The position of notches for pipes and holes for cables in joists should have been determined at the building design stage and indicated on the drawings, as both reduce the joist strength. Notches and holes in joists should be kept to a minimum and conform to the following (see Figures 2.39 and 2.40):

◆ Notches on the joist's top edge of up to 0.125 of the joist's depth located between 0.07 and 0.25 of the span from either support are permissible.
◆ Holes of up to 0.25 of the joist's depth drilled on the neutral stress line (centre line) and located between 0.25 and 0.4 of the span from either support are permissible. Adjacent holes should be separated by at least three times their diameter measured centre to centre.

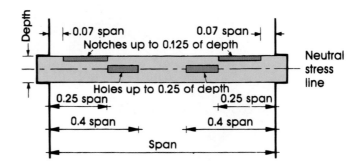

Figure 2.39 *Positions for notches and holes*

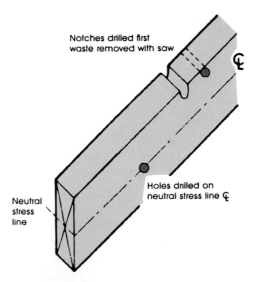

Figure 2.40 *Positioning of notches and holes to accommodate services*

example

The position and sizes for notches and holes in a 200 mm depth joist spanning 4000 mm are:

Notches – between 280 mm and 1000 mm in from each end of the joist and up to 25 mm deep.

Holes – between 1000 mm and 1600 mm in from each end of the joist and up to 50 mm diameter.

Excessive notching and drilling of holes outside the permissible limits will seriously weaken the joist and may lead to structural failure.

The plan of the building at upper floor joist bearing level is shown.

Internal walls are to receive a wall plate.

Joists are to be built into the external walls.

The flue is 500 mm square.

Double joists are required where partitions are to be built on the floor.

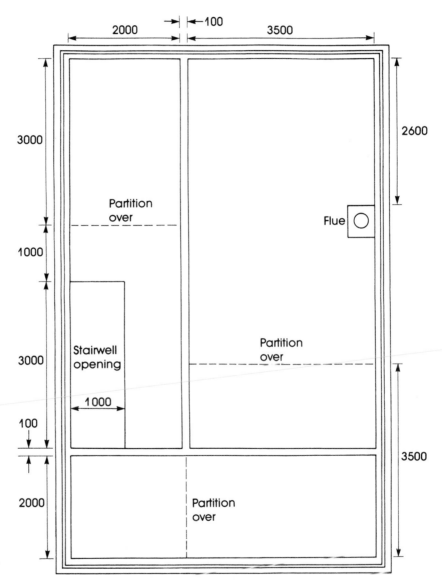

Figure 2.41 *Floor plan*

1. Determine and indicate on the drawing the position of all joists.
2. Name on the drawing all joists.
3. Indicate on the drawing any strutting required.
4. Produce a list of all materials required to complete the joist layout.

Carcassing **Chapter 2**

24. State the purpose of trimming joists.

25. Name the regulations that apply to the trimming of joists near to sources of heat.

26. What types of nails are used for fixing a thin galvanized steel joist hanger that connects a trimmed joist to a trimmer?

27. Use a sketch to define the neutral stress areas of a joist.

28. Produce a sketch to illustrate the proportions of a tusk mortise-and-tenon joint.

29. State the minimum distance for building in a joist adjacent to the flue of a heat-producing appliance.

30. Produce a sketch to illustrate the correct position/cutting of the end of a built in joist in relation to the cavity wall.

31. What is the reason for keeping the camber or crown of a joist upwards?

32. A joist has a large edge knot towards the centre of its span.

    (a)   Which way up should the joist be positioned?

    (b)   State the reason for your answer to (a).

33. State the reason why joists adjacent to walls are kept 50 mm away.

34. State the purpose of a trimmer joist used when forming openings.

35. State the purpose of notching and boring holes in joists.

36. Why are temporary battens sometimes fixed across the tops of joists?

37. State the maximum diameter hole that can be bored in a 200 mm deep joist.

38. Produce a sketch to show how a notch in the top of a joist could be formed to receive a 25 mm diameter pipe.

39. State the likely effect of excessive or wrongly placed notching to joists.

40. State why timber should not be used when packing joists to line in an external wall.

Joist coverings

Joist coverings provide the actual working surface and are variously termed 'flooring' or 'floorboarding' when used on a floor or as decking or boarding when used for flat roofs. The main materials in common use as illustrated in Figure 2.42 are:

◆ timber floorboards mainly PTG (planed, tongued and grooved) but square-edged boarding may be used for roofing;

◆ flooring grade particle board, mainly tongued and grooved or square-edged chipboard, but OSB (orientated strand board) is also used;

◆ flooring grade plywood either tongued and grooved or square-edged.

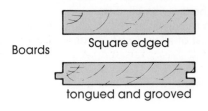

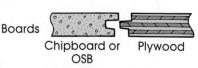

Figure 2.42 *Joist coverings*

Floorboarding is normally carried out after timber carcassing and preferably also after the window glazing and roof tiling is complete so that it is not exposed to the weather. Boarding or decking to flat roofs should be carried out just before they are to be waterproofed. Chipboard roof sheets are available pre-covered with a layer of felt to provide some measure of initial protection. Where sheet material is used in kitchens and bathrooms, a water-resisting grade should be used.

Hardwood strip flooring

Hardwood strip flooring is used where the floor is to be a decorative feature. It is made from narrow strips of tongued-and-grooved hardwood. These are usually random lengths up to 1.8m long and 50mm to 100mm in width. The heading joints are also usually tongued and grooved. Figure 2.43 shows a section through a piece of hardwood strip floorboard. It has a splayed-and-stepped tongue to ease secret nailing and also to reduce any likelihood of the tongue splitting.

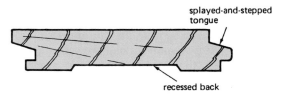

Figure 2.43 *Hardwood strip floorboards*

The purpose of the recessed back is to ensure positive contact between the board and joist. It is essential that the newly laid hardwood strip flooring is protected from possible damage by completely covering it with building paper or a polythene sheet, which should remain in position until the building is ready for occupation.

Softwood flooring

Softwood flooring usually consists of ex 25mm × 150mm tongued-and-grooved boarding.

A standard floorboard section has the tongue and groove offset away from the board's face. This identifies the upper face and also provides an increased wearing surface before exposing the tongue and groove. Boards can be fixed either by floor brads nailed through the surface of the boards and punched in, or by lost-head nails secret fixed through the tongue as shown in Figure 2.44. Nails should be approximately 2½ times the thickness of the floorboard in length.

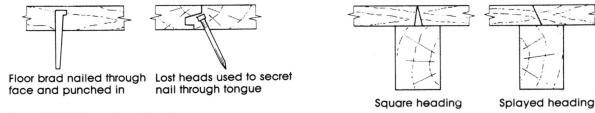

Floor brad nailed through face and punched in   Lost heads used to secret nail through tongue

Figure 2.44 *Fixing softwood flooring*

Square heading   Splayed heading

Figure 2.45 *Heading joints for softwood flooring*

Figure 2.45 shows square-butted or splayed heading joints, which are introduced as required to utilize offcuts of board and avoid wastage. Splayed heading joints are preferred as there is less risk of the board end splitting. Heading joints should be staggered evenly throughout the floor for strength; these should never be placed next to each other, as the joists and covering would not be tied together properly as illustrated in Figure 2.46.

Carcassing **Chapter 2**

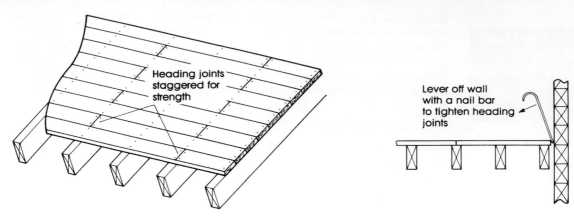

Figure 2.46 Positioning of heading joints

Surface fixing

Boards are laid at right angles to the joists. The first board should be fixed at least 10 mm away from the outside wall. This gap, which is later covered by the skirting, helps to prevent dampness being absorbed through direct wall contact. In addition, the gap also allows the covering material to expand without either causing pressure on the wall or a bulging of the floor surface. The remainder of the boards are laid four to six at a time, cramped up with floorboard cramps (see Figure 2.47) and surface nailed to the joists. The final nailing of a floor is often termed 'bumping'. This should be followed by punching the nail head just below the surface. Boards up to 100 mm in width require one nail to each joist while boards over this width require two nails.

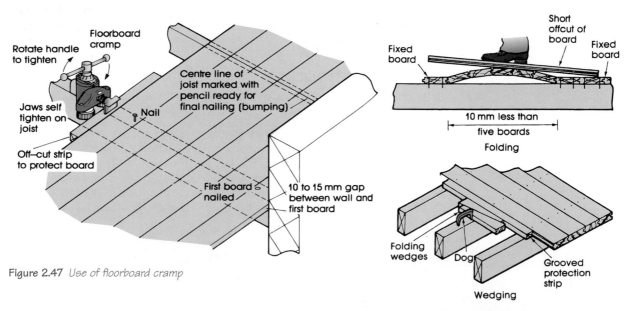

Figure 2.47 Use of floorboard cramp

Figure 2.48 Alternative methods of tightening surface-fixed floorboards

did you know?

A gap should be maintained between the outer floorboards and the wall. This prevents the possibility of dampness and allows for expansion.

As Figure 2.48 shows two alternatives to the use of floorboard cramps are folding or wedging, although neither of these is as quick or efficient. Folding a floor entails fixing two boards spaced apart 10 mm less than the width of five boards. The five boards can then be placed with their tongues and grooves engaged. A short board is laid across the centre and 'jumped on' to press the boards in position. This process is then repeated across the rest of the floor. Alternatively, the boards may be cramped, four to six at a time, using dogs and wedges.

Secret fixing

Secret fixed boards must be laid and tightened individually and cramping is therefore not practical. Figure 2.49 shows how they may be tightened by levering them forward with a firmer chisel driven into the top of the joist, or with the aid of a floorboard nailer. This tightens the boards and drives the nail when the plunger is struck with a hard mallet.

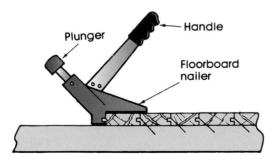
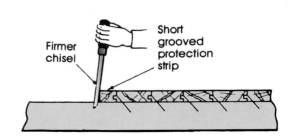

Figure 2.49 *Tightening secret fixed floorboards*

Secret fixing is normally only used on high-class work or hardwood flooring, as the increased laying time makes it considerably more expensive.

Services

safety tip

Take extra care near services! Don't nail a pipe or a cable.

Where services such as water and gas pipes or electric cables are run within the floor, there is a danger of driving nails into them. They should be marked in chalk or pencil 'PIPES NO FIXING', so that on nailing the danger area is kept clear (see Figure 2.50). The floorboards over services can be fixed with recessed cups and screws to permit easy removal for subsequent access, and also to provide easy recognition of location.

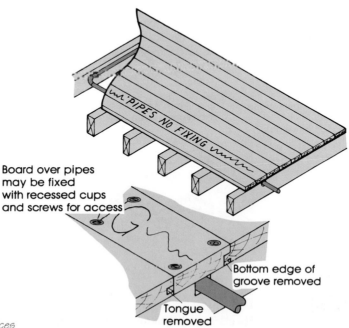

Figure 2.50 *Marking position of services*

Access traps, Figure 2.45, may be required in a floor over areas where water stop cocks or electrical junction boxes are located. Again these can be fixed with recessed cups and screws to permit easy removal.

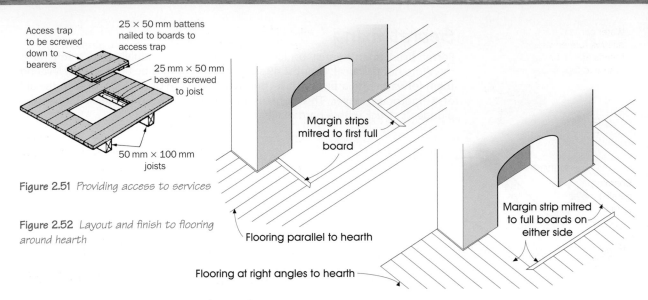

Figure 2.51 *Providing access to services*

Figure 2.52 *Layout and finish to flooring around hearth*

Fireplace openings

Figure 2.52 shows how margin strips are used to finish the ends of the floorboarding around a fire hearth. The actual arrangement will vary depending on whether the boards run into or across the hearth. In either case the layout of boarding should be planned before any fixing takes place, to ensure an even margin around the hearth.

Sheet floor decking

Sheet floor decking is now being increasingly used for domestic flooring. Flooring-grade chipboard is available with square edges in 1220 mm × 2440 mm sheets, and with tongued and grooved edges in 600 mm × 2440 mm sheets. Square-edged sheets are normally laid with their long edges over a joist. Noggins must be fixed between the joists to support the short ends. Tongued-and-grooved sheets are usually laid with their long edges at right angles to the joists and their short edges joining over the joist. Both types require noggins between the joists where the sheet abuts a wall. See Figure 2.53. Joists should be spaced to accommodate the dimensions of the sheet flooring.

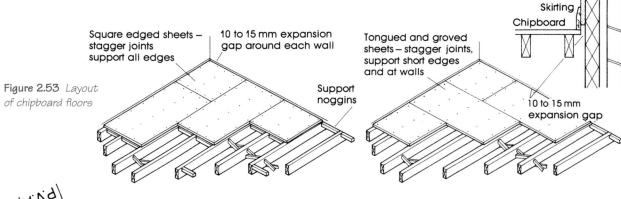

Figure 2.53 *Layout of chipboard floors*

Figure 2.54 *Gluing joints in chipboard flooring*

Fixing

Sheets are laid staggered and fixed at 200 mm to 300 mm centres with 50 mm lost-head nails or, for additional strength, annular ring shanked or serrated nails. A gap of 10 mm must be left along each wall to allow for expansion and prevent absorption of dampness from the wall. Manufacturers of chipboard often recommend the gluing of the tongues and grooves with a PVA adhesive to prevent joint movement and stiffen the floor. See Figure 2.54.

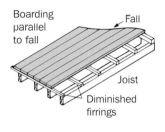

Water can collect in the hollows formed when boards are laid at right angles to fall

Fall

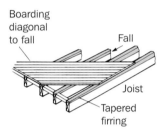

Boarding parallel to fall

Fall

Joist

Diminished firrings

Boarding diagonal to fall

Fall

Joist

Tapered firring

Figure 2.55 *Timber boards for flat roof decking*

For protection it is recommended that the floor be covered with building paper after laying and that this is left in position until the building is occupied.

Plywood flooring is laid using the same procedures as chipboard flooring.

Flat roof decking

The choice of materials and method of laying decking can be likened to floors with the exception that where timber boards are used they should be laid either with, or at a diagonal to, the fall of the roof's surface. Cupping or distortion of the boards can lead to pools of water being trapped in hollows formed on the roof's surface, where boards are laid at right angles to the fall. See Figure 2.55.

Determining materials

Simple calculations are used in order to determine the amount of joist covering materials required for an area.

To determine the area of a simple rectangular room multiply its width by its length.

example

Calculate the floor area of a room 3.6 m wide by 4.85 m long.

Area of floor $= 3.6 \times 4.85$

$= 17.46\,\text{m}^2$

To determine more complex floor areas, divide them into a number of rectangles or other recognizable shapes and work out the area of each in turn.

example

Calculate the floor area of the room shown in Figure 2.56. This can be divided into two rectangles, A and B, each being solved separately then added together. (As an area in square metres, m^2, is required change all dimensions to metres before starting.)

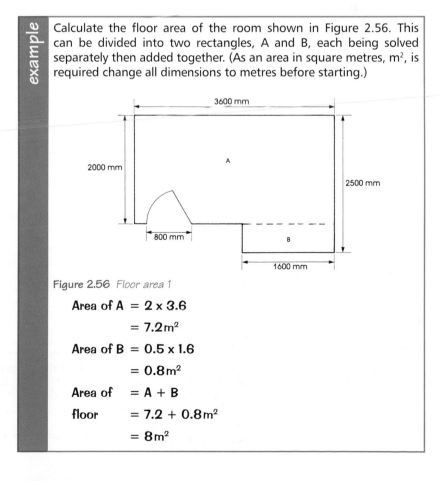

Figure 2.56 *Floor area 1*

Area of A $= 2 \times 3.6$

$= 7.2\,\text{m}^2$

Area of B $= 0.5 \times 1.6$

$= 0.8\,\text{m}^2$

Area of $= A + B$

floor $= 7.2 + 0.8\,\text{m}^2$

$= 8\,\text{m}^2$

example

The area of the room shown in Figure 2.57 is equal to area A plus area B minus area C.

Area A = $(9 + 10.5) \div 2 \times 6.75$
(Trapezium) = $65.813\,\text{m}^2$

Area B = 0.75×5.5
= $4.125\,\text{m}^2$

Area C = 0.9×3
= $2.7\,\text{m}^2$

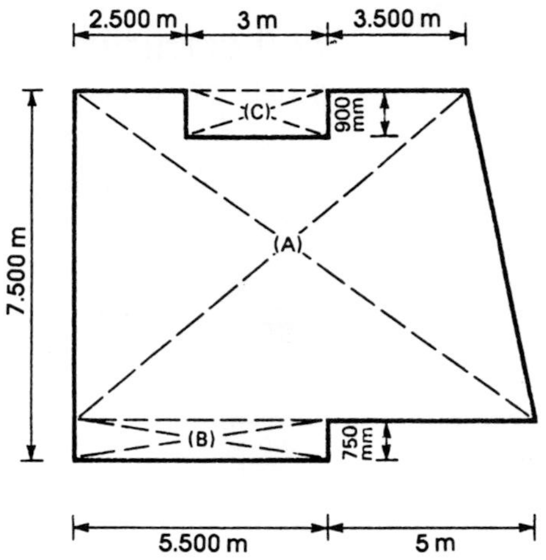

Figure 2.57 *Floor area 2*

Total area of floor = $A + B - C$
= $65.813 + 4.125 - 2.7$
= $67.238\,\text{m}^2$

In order to determine the metres run of floorboards required to cover a room, the floor area is divided by the board's covering width.

example

Calculate the metres run of floorboards required to cover a floor area of 4.65 m², if the floorboards have a covering width of 137 mm. (Change 137 mm to metres before starting by moving its imaginary point (behind the 7) three places forward to become 0.137 m. This is because all the units in a calculation must be the same.)

Metres run required = Area ÷ Width of board

= $4.65 \div 0.137$

= $33.94\,\text{m}$
say **34 m run.**

It is standard practice to order an additional amount of flooring to allow for cutting and wastage. This is often between 10% and 15%.

example

To calculate the metres run of floor boarding required plus an additional percentage, turn the percentage into a decimal:

e.g. 5% = 0.05; 10% = 0.1; 25% = 0.25

Place a 1 in front of the point (to include original amount) and use this number to multiply the original amount, e.g. for 5% increase use 1.05 to multiply, for 10% increase use 1.1, for 25% increase use 1.25.

> **example**
>
> If 34m run of floorboarding is required to cover an area, calculate the amount to be ordered including an additional 12% for cutting and wastage. (For 12% increase multiply by 1.12.)
>
> **Amount to be ordered = 34 x 1.12**
>
> **= 38.08**
>
> **say 38m run.**

In order to determine the number of sheets of plywood or chipboard required to cover a room either:

◆ divide area of room by area of sheet; or
◆ divide width of room by width of sheet, divide length of room by length of sheet.

Convert these numbers to the nearest whole or half and multiply them together.

> **example**
>
> Calculate the number of 600mm × 2400mm chipboard sheets required to cover a floor area of 2.05 × 3 3.6m.
>
> **Area of room   = 2.05 x 3.6**
> **   = 7.38m²**
> **Area of sheet   = 0.6 x 2.4**
> **   = 1.44m²**
> **Number of sheets required   = Area of room ÷ Area of sheet**
> **   = 7.38 ÷ 1.44**
> **   = 5.125**
> **   say 6 sheets**
> **   or alternatively**
> **Number of sheet widths in room width   = 2.05 ÷ 0.6**
> **   = 3.417**
> **   say 3.5**
> **Number of sheet lengths in room   = 3.6 ÷ 2.4**
> **length   = 1.5**
> **Total number of sheets   = 3.5 x 1.5**
> **   = 5.25**
> **   say 6 sheets**

activity

The bungalow shown in Figure 2.58 is to have a timber suspended hollow ground floor to all rooms except the garage. The overall internal measurements are 10,950 mm × 6650 mm, the garage is 2750 mm × 4550 mm.

Figure 2.58 *Bungalow plan*

Determine the overall floor area in square metres (including the garage).

Determine the required area of joist covering in metres square.

Calculate the amount of T & G boarding required in metres run; allow 15% extra as a cutting allowance.

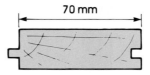

70 mm

Figure 2.59 *Boarding 70 mm*

41. Produce a sketch to show the section of a tongued and grooved floorboard.

42. State the purpose of heading joints in timber boarded floor surfaces and explain why they should be staggered.

43. State ONE reason why an access trap may be required in joist covering.

44. State the reason why a gap is left between joist coverings and adjacent walls.

45. Name a nail, stating its length, which is suitable for the surface fixing of a 20 mm finished thickness floorboard.

46. Explain ONE method that can be used to cramp up tongued and grooved floorboards in the absence of flooring cramps.

47. Explain the purpose of noggins when using sheet joist coverings.

48. Explain the purpose of gluing the tongues and grooves of sheet joist coverings.

49. Name one nail suitable for fixing sheet joist coverings.

50. Produce a sketch to show what is meant by secret fixing when applied to tongued and grooved floorboards.

Roof types

Roofs are the uppermost part of a building. They span the external walls and provide protection from the elements. They may be classified according to their pitch, shape and also type of structure.

Roof pitch

The pitch, slope or inclination of a roof surface to the horizontal may be expressed either in degrees or as a ratio of the rise to the span, see Figure 2.60.

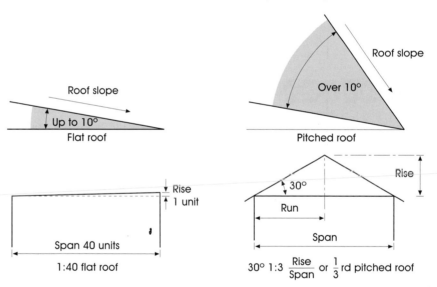

Figure 2.60 *Roof pitch*

A flat roof with a slope of 1:40 will rise 1 unit for every 40 units run, that is a rise of 25 mm for every metre run; a one-third pitch roof (1/3 or 1:3) with a span of 3 m will have a rise of 1 m.

Pitch is dependent on the type and size of the materials used for the weatherproof covering and the building's exposure to windy conditions. They may be classified as either:

◆ flat roofs where the pitch of the roof surface does not exceed 10°; or
◆ pitched roofs where the pitch of the roof surface exceeds 10°.

Flat roofs are normally weathered using either rolls of bituminous felt built up in layers and bonded to the decking with cold or hot applied bitumen; or mastic asphalt hot applied in layers on a sheathing felt that covers the decking. In order to

Carcassing **Chapter 2**

reflect some of the heat of the sun, both types of weathering may be protected with a heat-reflecting paint or covered with a layer of pale-coloured stone chippings.

Pitched roofs can be covered using clay or concrete tiles; quarried or man-made slates; wood shingles (normally cedar); thatch (reeds); and various profiled sheet materials manufactured from protected metals, plastics and cement/mineral fibre mix. In general, as the pitch of the roof lowers the unit size of the covering material must increase.

Roof shape

Roofs may be further divided into a number of basic forms by their shape, the most common of which are illustrated in Figure 2.61.

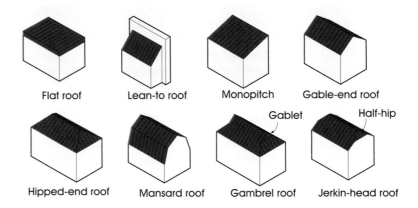

Figure 2.61 *Types of roof*

safety tip

When erecting a roof you will be working at heights. Always use an appropriately guarded working platform and a properly secured ladder to access it.

Lean-to and mono-pitch roofs

Both lean-to and mono-pitch roofs have a single sloping surface. The lean-to roof abuts a higher wall or building and is often used for extensions, whereas mono-pitched roofs are normally freestanding.

Gable-end roofs

Gable-end roofs have two sloping surfaces (double pitched) terminating at one or both ends with a triangular section of brickwork.

Hipped-end roofs

Hipped-end roofs are again double-pitched, but in these cases the roof slope is returned around one or both the shorter sides of the building to form a sloping triangular end.

Mansard roofs

Mansard roofs are double-pitched roofs, where each slope has two pitches. The lower part has a steep pitch (to act as walls) and often incorporates dormer windows for room in the roof applications while the upper part (acting as the main weathering) rarely exceeds 30°. The shorter sides of the building may be finished with either gable or hipped ends.

Gambrel roofs

Gambrel roofs are double-pitched roofs incorporating a small gable or gablet at the ridge and a half-hip below. The gablet may be finished with tile hanging, timber cladding or a louvred ventilator.

Jerkin-head roofs

Jerkin-head roofs are double-pitched roofs, which are hipped from the ridge part-way to the eaves, and the remainder gabled. These are also known as Dutch hipped roofs.

Roof elements

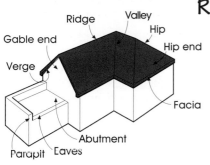

Figure 2.62 *Roof elements*

- ◆ **Abutment** – *the intersection where a lean-to or flat roof meets the main wall structure.*
- ◆ **Eaves** – *the lower edge of a roof, which overhangs the walls and where rainwater is discharged normally into a gutter. May be finished with a fascia board and soffit.*
- ◆ **Gable** – *the triangular upper section of a wall that closes the end of a building with a pitched roof.*
- ◆ **Hip** – *the line between the ridge and eaves of a pitched roof, where the two sloping surfaces meet at an external angle.*
- ◆ **Parapet** – *the section of a wall that projects above and is terminated some way beyond the roof surface.*
- ◆ **Ridge** – *the horizontal line of intersection between two sloping roof surfaces at their highest point or apex.*
- ◆ **Valley** – *the line between the ridge and eaves of a pitched roof, where the two sloping surfaces meet at an internal angle.*
- ◆ **Verge** – *the termination or edge of a pitched roof at the gable or the sloping edge (non-drained) of a flat roof. May be finished with a bargeboard and soffit.*

Roof construction

Timber flat roofs

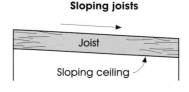

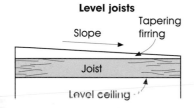

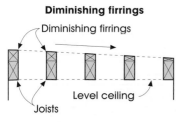

Figure 2.63 *Methods of forming flat roof slopes*

Timber flat roofs are similar in construction to that of upper floors. Their means of support, tying in/down, stiffening with the use of strutting and the decking all follow on similar lines.

Although the surface of the roof is flat, it is not horizontal. It should have a slope or fall on its top surface to ensure that rainwater will be quickly cleared and not accumulate on the roof. There are three ways of forming the slope as illustrated in Figure 2.63:

1. Level joists – the joists are laid level with the slope being formed by long tapering wedged shaped pieces of timber (firring pieces) being nailed to their upper edges. This is the most common method as it forms a level under surface for fixing the ceiling plasterboard.

2. Sloping joists – the joists themselves are laid to the required slope and therefore no firrings are required. However it has the disadvantage of forming a sloping ceiling soffit.

3. Diminishing firrings – the joists are laid level at right angles to the slope and reducing or diminishing sections of timber are nailed to their top edge to create the fall. This method is rarely seen, but it is useful when timber boards are used for the roof decking, as these should be fixed parallel to the roof slope.

Flat roof carcass

Joist size and spacing

The size of flat roof joists can be worked out in the same way as floor joists. For new work the sectional size and spacing will always be specified or alternatively standard tables can be used. Slightly longer spans are permitted for flat roof joists than floor joists of the same sectional size, as flat roofs are only normally walked upon for maintenance purposes. The actual spacing of the joists must be related to the type of material used for the decking. Table 2.3 can be used as a guide.

Table 2.3 *Maximum clear span of flat roof joists and decking (m)*

Based on the use of C16 grade softwood with dead load of up to 0.5 kN/m² with access only for maintenance and repair purposes.

Size of joist (mm)	Joist spacing (centre/centre)		
B × D	400 mm	450 mm	600 mm
38 × 97	1.74	1.56	1.21
38 × 122	2.37	2.22	1.76
38 × 147	3.02	2.71	2.33
38 × 170	3.63	3.10	2.69
38 × 195	4.30	3.52	3.06
38 × 220	4.94	3.93	3.42
50 × 97	1.97	1.87	1.54
50 × 122	2.67	2.50	2.19
50 × 147	3.39	3.01	2.69
50 × 170	4.06	3.47	3.08
50 × 195	4.79	3.97	3.50
50 × 220	5.38	4.47	3.90
75 × 122	3.17	2.86	2.60
75 × 147	3.98	3.43	3.13
75 × 170	4.74	3.96	3.61
75 × 195	5.42	4.52	4.13
75 × 220	6.07	4.97	4.64

Decking material	Finished thickness	Span (joist spacing)
Softwood T & G	16 mm	up to 500 mm
Softwood T & G	19 mm	up to 600 mm
Roofing grade chipboard	18 mm	up to 500 mm
Roofing grade chipboard	22 mm	up to 600 mm
Orientated strand board OSB	15 mm	up to 500 mm
Orientated strand board OSB	18 mm	up to 600 mm
External plywood WBP	16 mm	up to 500 mm
External plywood WBP	19 mm	up to 600 mm
Wood wool Slabs	50 mm	up to 600 mm

did you know?

Recommended span of decking may vary between manufacturers of panel products. Thus always check the specific manufacturer's details.

Layout of joists

The positioning and levelling of flat roof joists is similar to the positioning and levelling of floor joists, except that flat roofs only occasionally have to be trimmed around openings, e.g. roof lights, chimney stacks etc. Flat roofs, like floors, also require some form of strutting at their mid span, see Figure 2.64.

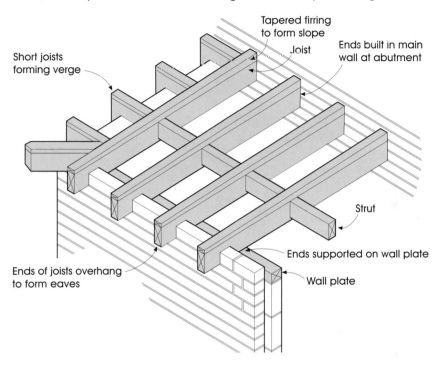

Figure 2.64 *Flat roof carcass*

Joist supports

The joists are supported at the eaves and/or verge depending on the joist's direction of span, either directly on the inner leaf of the cavity wall or a timber wall plate that has been secured on top of the inner leaf (see Figure 2.65). Where the ends of the joists abut a wall either at the main building or at a parapet various methods can be used for support, including building-in, joist hangers, timber wall pieces and steel angle sections.

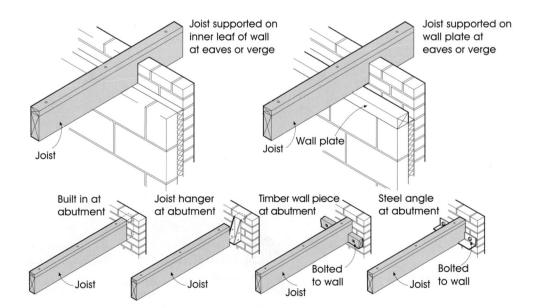

Figure 2.65 *Flat roof joist supports*

Anchoring joists

In order to prevent strong winds lifting the roof, joists must be anchored to the walls at a maximum of 2 metre centres. This anchoring must be carried out to the joists that run both parallel with or at right angles to the wall. Figure 2.66 shows how a flat roof joist can be anchored to the wall using an anchoring or tie-down strap. The strap is nailed to the side of the joist and fixed to the wall by either nailing or plugging and screwing.

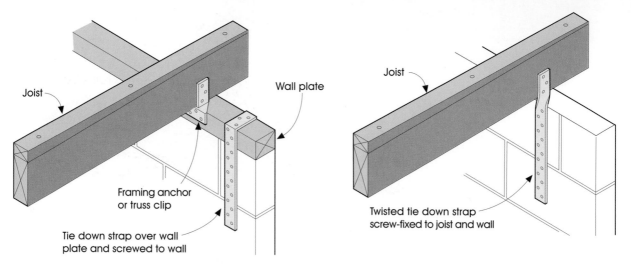

Figure 2.66 *Anchoring joists*

Single or double flat roofs

Binders may be incorporated in large flat roofs, either to provide intermediate support to long joists or to enable the roof to slope in both directions as illustrated in Figure 2.67. The use of a binder creates what is termed as a double flat roof. This reduces the effective span of the joists and enables the use of smaller sectional sizes. In addition the required width of verge and eave finishes are reduced.

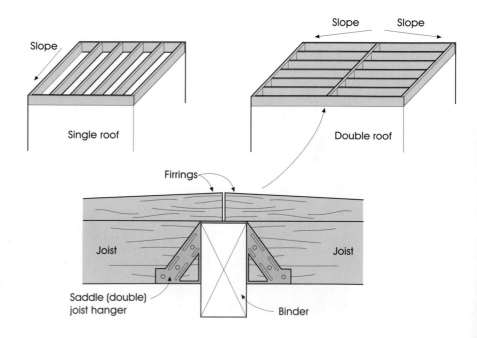

Figure 2.67 *Single and double flat roofs*

Timber pitched roofs

Timber pitched roofs may be divided into two broad but distinct categories:

◆ *Traditional framed cut roofs* – almost entirely constructed on-site from loose timber sections and utilizing simple jointing methods (see Figure 2.68).
◆ *Prefabricated trussed rafters* – normally manufactured under factory conditions, from prepared timber butt-jointed and secured using nail plates (see Figure 2.69).

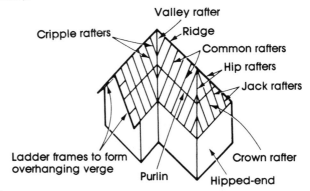

Figure 2.68 *Terminology for traditional framed cut roof*

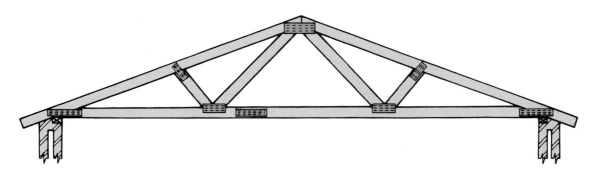

Figure 2.69 *Trussed rafter*

Traditional roofs

Traditional roofs may be constructed as either single or double roofs, according to their span.

Single roofs

The rafters of single roofs do not require any intermediate support (see Figure 2.70). They are not economically viable when the span of a roof exceeds about 5.5 m. This is because very large sectioned timber would have to be used. A central binder may be hung from the ridge to bind together and prevent any sagging when ceiling joists are used in a close coupled single roof.

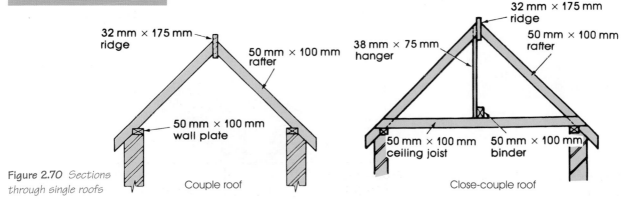

Figure 2.70 *Sections through single roofs*

Carcassing **Chapter 2**

Double roofs

The rafters of double roofs are of such a length that they require an intermediate support (see Figure 2.71). This is normally given by purlins to support the rafters in mid span.

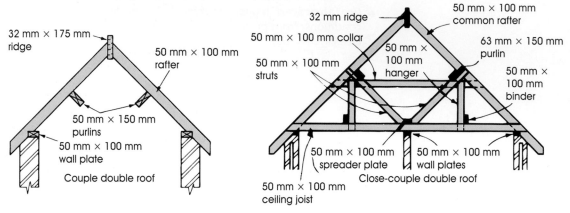

Figure 2.71 *Sections through double roofs*

The use of purlins effectively reduces the span of the rafters and limits their size to economical sections. Struts, collars and hangers are spaced evenly along the roof every third or forth pair of rafters, to provide intermediate support for the purlins. Binders may also be fixed to the ceiling joists and hung from above to prevent any sagging of the ceiling.

Cut roofs component terminology

Cut roofs may be constructed as either single or double roofs, according to their span.

◆ **Common rafters** – the main load-bearing timbers in a roof, which are cut to fit the ridge and birdsmouthed over the wall plate (see Figure 2.72).
◆ **Ridge** – the backbone of the roof, which provides a fixing point for the tops of the rafters, keeping them in line.

did you know?

Double roofs incorporate purlins, which are timber beams that are positioned at right-angles under the common rafters to provide a support in their mid-span. Purlins themselves may be supported at intervals by timber struts.

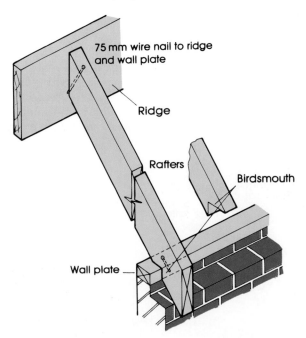

Figure 2.72 *Fixing a common rafter*

- **Purlin** – a beam that provides support for the rafters in their mid span (see Figures 2.73 and 2.74).
- **Ladder frame** – this is also known as a gable ladder and is fixed to the last common rafter to form the overhanging verge on a gable roof. It consists of two rafters with noggins nailed between them (see Figure 2.74).
- **Wall plate** – this transfers the loads imposed on the roof, uniformly over the supporting brickwork. It also provides a bearing and fixing point for the feet of the rafters.
- **Ceiling joists** – as well as being joists on which the ceiling is fixed, they also act as ties for each pair of rafters at wall plate level.
- **Binders and hangers** – these stiffen and support the ceiling joists in their mid span, to prevent them from sagging and distorting the ceiling.
- **Struts** – transfer the loads imposed by the purlins onto a load-bearing partition wall.
- **Spreader plate** – provides a suitable bearing for the struts at ceiling level.
- **Collar tie** – prevents the spread of rafters in the same way as ceiling joists and may also be used to provide some support for the purlins.

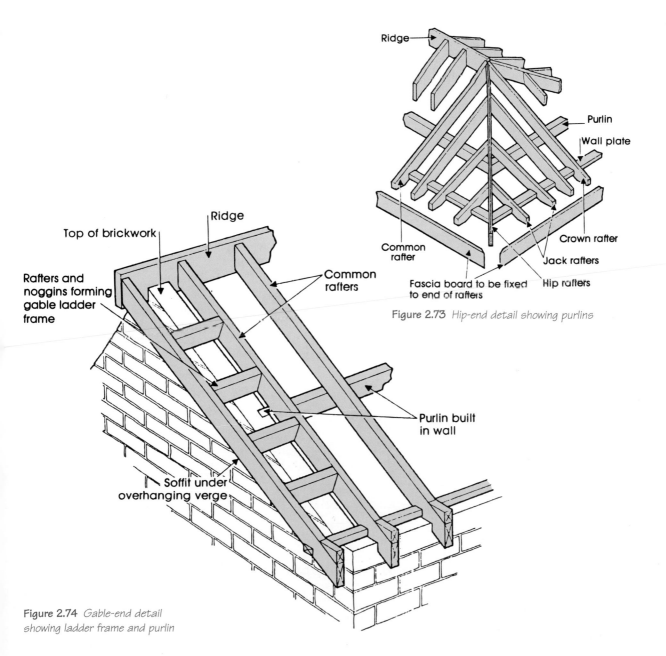

Figure 2.73 Hip-end detail showing purlins

Figure 2.74 Gable-end detail showing ladder frame and purlin

Trimming

Where openings occur in roofs, in either the rafters or ceiling joists or both, these have to be trimmed (see Figures 2.75 and 2.76). Framing anchors or housing joints are used to join the trimmers, trimmings and trimmed components together.

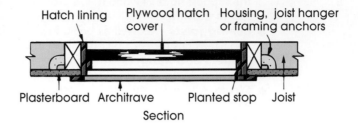

Figure 2.75 *Trimming to loft hatch*

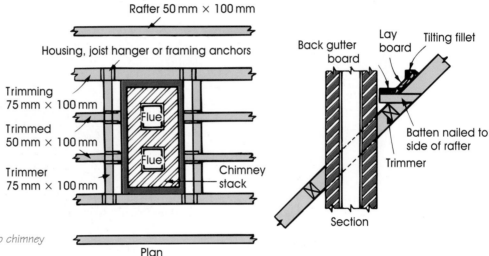

Figure 2.76 *Trimming to chimney stack*

When trimming around a chimney stack, in order to comply with Building Regulations, no combustible material, including timber, is to be placed within 200 mm of the inside of the flue lining; or, where the thickness of the chimney surrounding the flue is less than 200 mm, no combustible material must be placed within 40 mm of the chimney.

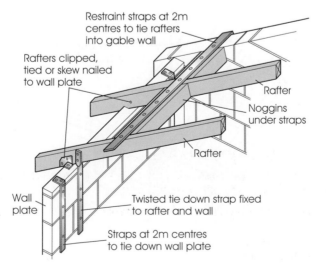

Figure 2.77
Anchoring roofs

Anchoring roofs

Rafters can be skew nailed to the wall plates or, in areas noted for high winds, framing anchors or truss clips where appropriate can be used (see Figure 2.77).

Wall plates must be secured to the wall with straps at 2 m centres.

The rafters and ceiling joists adjacent to the gable end should also be tied into the wall with metal restraint straps at 2 m centres.

Water-tank platforms

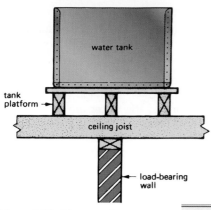

Figure 2.78 *Water-tank platform*

These are ideally situated centrally over a load-bearing wall. The platform should consist of boarding supported by joists laid on top of the ceiling joists and at right angles to them. Figure 2.78 illustrates a typical water-tank platform.

Roof sizes and spacing

Sizes and spacing of members are dependent on the location, pitch and span of the roof. For new work, the sectional size and spacing to use will always be specified. In other situations, Table 2.4 can be used as a guide.

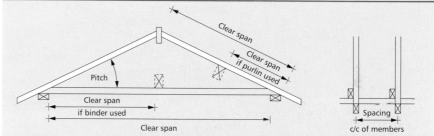

Size of joist (mm)	Joist spacing (centre/centre)			
B × D	400 mm	450 mm	600 mm	
38 × 100	2.28	2.23	2.10	Based on using C16 grade softwood rafters for roofs over 30° and up to 45° pitch, with a dead load of up to 0.5 kN/m^2, imposed loading of 0.75 kN/m^2 suitable for most sites located up to 100 m above ordnance datum with access only for maintenance and repair. Ceiling joists are for dead loads between 0.25 and 0.50 kN/m^2
38 × 125	3.07	2.95	2.69	
38 × 150	3.67	3.53	3.22	
50 × 100	2.69	2.59	2.36	
50 × 125	3.35	3.23	2.94	
50 × 150	4.00	3.86	3.52	
Size of ceiling joist (mm)				
38 × 72	1.11	1.10	1.06	
38 × 97	1.67	1.64	1.58	
38 × 122	2.25	2.21	2.11	
38 × 147	2.85	2.80	2.66	
50 × 72	1.27	1.25	1.21	
50 × 97	1.89	1.86	1.78	
50 × 122	2.53	2.49	2.37	
50 × 147	3.19	3.13	2.97	

Table 2.4 *Maximum clear span of pitched roof rafters and ceiling joists (m)*

Determining lengths and bevels

There are three main methods that can be used to determine the lengths and bevels required for a traditional cut roof:

◆ full-size setting out, only really suitable for fairly small rise/span gable end roofs;
◆ scale drawings;
◆ roofing square.

Full size setting out

This method involves setting out in full size on a suitable surface, such as a concrete base, a couple of sheets of plywood, or couple of scaffold boards or joists positioned at right angles to each other.

For practical purposes the initial setting out is done using single lines, without the width or thickness of the members being shown, see Figure 2.79. Lines to indicate the rise and run (half the roof span) of the rafter are drawn first, followed by the diagonal rafter line (pitch line). This will indicate the required plumb and seat cuts as well as the theoretical or true length of the rafter. The end sections of the wall plate and the ridge can be added and then the width of the rafter, two-thirds above and one-third below the pitch line.

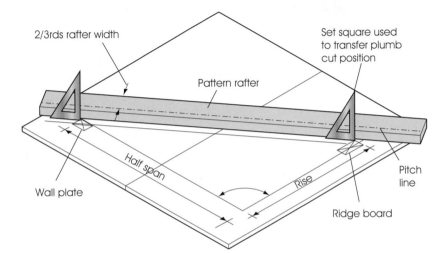

Figure 2.79 *Marking pattern rafter from full size setting out*

A pattern rafter can then be laid on top of this and marked out. The rest of the rafters can then be marked out and cut from the pattern.

Scale drawings

After having produced your own scale drawings or having had a setter-out produce them, either manually or with the aid of a computer-drawing programme (CAD), you will have to interpret the information in order to mark out the roof members.

Figure 2.80 shows a typical scale drawing of the section through a pitched roof. The drawing has been marked up to show the angles and true length of the common rafter.

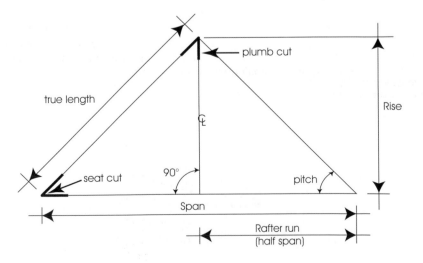

Figure 2.80 *Scale drawing showing section through pitched roof*

An adjustable bevel square can be set up to the required angles by placing over the actual drawing as shown in Figure 2.81.

The true length shown on the drawing is measured on the pitch line, which is a line marked up from the underside of the common rafter, one-third of its depth, as shown in Figure 2.82.

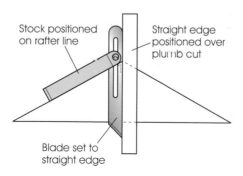

Stock positioned on rafter line

Straight edge positioned over plumb cut

Blade set to straight edge

Figure 2.81 *Setting an adjustable bevel for a rafter plumb cut*

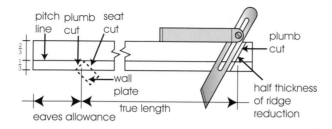

pitch line plumb cut seat cut

$\frac{2}{3}$ $\frac{1}{3}$

wall plate

eaves allowance

true length

plumb cut

half thickness of ridge reduction

Figure 2.82 *Setting out a common rafter*

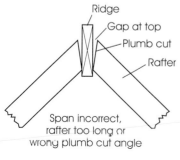

Ridge
Gap at top
Plumb cut
Rafter

Span incorrect, rafter too long or wrong plumb cut angle

Gap at bottom

Span incorrect, rafter too short or wrong plumb cut angle

Figure 2.83 *Possible effect of scaling errors*

Take care when scaling the dimensions from the drawing, as 1 mm difference on a 1:20 drawing is 20 mm difference at full size. It is best always to cut a pair of rafters and try them up on the roof for size, before progressing onto cutting the full requirement. Figure 2.83 illustrates the possible effect of any scaling errors in the length of rafters.

The true length of the common rafter is measured on the pitch line from the centre line of the ridge to the outside edge of the wall.

Therefore, when marking out the true length of the common rafter from the single line drawing, an allowance in measurement must be made. This allowance should be an addition for the eaves overhang and a reduction of half the ridge thickness.

Roofing square

The wide part of a roofing square is the blade and the narrow part is the tongue (see Figure 2.84). Both the blade and the tongue are marked out in millimetres. Most carpenters will make a fence for themselves using two battens and four small bolts and wing nuts.

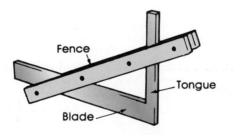

Fence

Tongue

Blade

Figure 2.84 *Roofing square and fence*

Carcassing **Chapter 2**

To set out a roof using the roofing square, the rise of the roof is set on the tongue and the run of the rafter is set on the blade (run of rafter = half of the span). In order to set the rise and run on the roofing square, these measurements must be scaled down and it is usual to divide them by 10.

did you know?

The roofing square, like the geometrical method, gives the true lengths of members on the pitch line. Therefore the same allowances in measurement as stated before must be made.

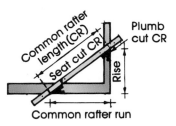

Figure 2.85 *Roofing square set-up for common rafter*

example

For a roof with a rise of 2.5 m and a rafter run of 3.5 m, the scale lengths to set on the roofing square would be:

$$\text{Rise } \frac{2.5}{10}\text{m} = 250\,\text{mm}; \text{ run } \frac{3.5}{10}\text{m} = 350\,\text{mm}$$

Figure 2.85 shows how to set up the roofing square and fence to obtain the required length and angles. The length will, however, be the scale length.

Figure 2.86 shows how the roofing square may be stepped down the rafter 10 times to obtain its actual length. Alternatively, the scale length can be measured off the roofing square and multiplied by ten to give its actual length.

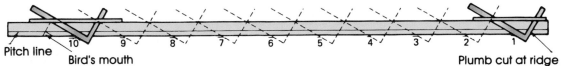

Figure 2.86 *Using a roofing square*

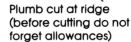
Cut roof erection

Gable roof

The procedure of erection for a gable-end single roof is as follows (see Figure 2.87):

safety tip

Roof erection is at least a two-person job. Never try to do it on your own!

1. The wall plate, having been bedded and levelled by the bricklayer, must be tied down.
2. Mark out the position of the rafters on the wall plate and transfer these onto the ridge.
3. Place ceiling joists in position and fix by nailing to the wall plate; placing scaffold boards over the ceiling joists can create a temporary working platform inside the roof space.
4. Make up two temporary A-frames. These each consist of two common rafters with a temporary tie joining them at the top, leaving a space for the ridge and a temporary tie nailed to them in the position of the ceiling joists for ease of handling.
5. Stand up the A-frames at either end of the roof, fix in position by skew nailing to the wall plate and the ceiling joists.
6. Fix temporary braces to hold the A-frames upright.
7. Insert the marked out ridge board and nail in position through the top edge of the A-frame rafters.
8. Progressively position pairs of rafters along the roof, securing them by nailing them at both the ridge and wall plate. Diagonal rafter bracing may be required to stabilize the roof in windy locations.
9. Add noggins to form a gable ladder where an overhanging verge is required.

Chapter 2 Carcassing

10. Allow bricklayers to build the gable-end walls.
11. Finish the roof at the verge and eaves with bargeboard, fascia and soffit as required. (This stage is covered later in this chapter.)

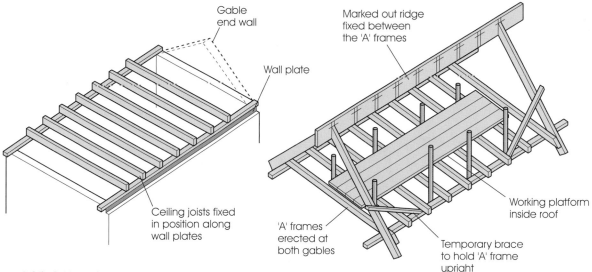

Figure 2.87 *Gable roof erection*

The erection of double-cut roofs follows the same procedure, except for the positioning and fixing of the purlin and any associated struts and binders. These may be fitted either after the A-frames and a central pair of rafters have been fixed, or left until all the common rafters have been fixed. The former method is often preferred, especially for longer spans, as an early fixed purlin provides a point on which to slide the intermediate common rafters up to the ridge and eases any potential problems associated with fixing to a run of possibly sagging rafters.

 # Prefabricated trussed rafters

Trussed rafters (Figure 2.88) are prefabricated by a number of specialist manufacturers in a wide range of shapes and sizes. They consist of prepared timber laid out in one plane, with their butt joints fastened with nail plates.

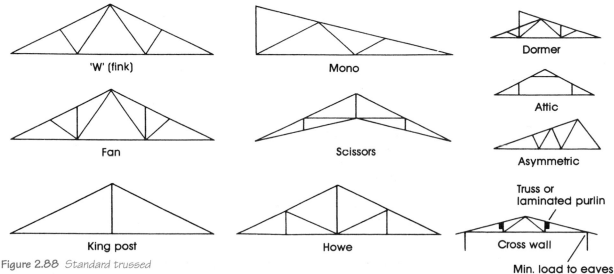

Figure 2.88 *Standard trussed rafter configurations*

In use the trusses are spaced along the roof at between 400 mm and 600 mm centres and fixed to the wall plate, preferably using truss clips (Figure 2.89).

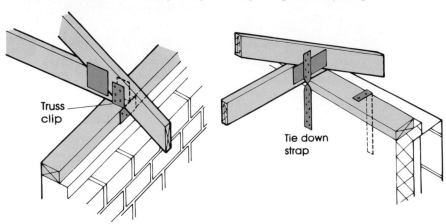

Figure 2.89 *Fixing trussed rafters to wall plate*

In order to provide lateral stability the roof requires binders at both ceiling and apex level and diagonal rafter bracing fixed to the underside of the rafters (see Figure 2.90). These must be fixed in accordance with the individual manufacturer's instructions.

Figures 2.91 and 2.92 show how the gable wall must be tied back to the roof for support. This is done using lateral restraint straps at 2 m maximum centres both up the rafter slope and along the ceiling tie.

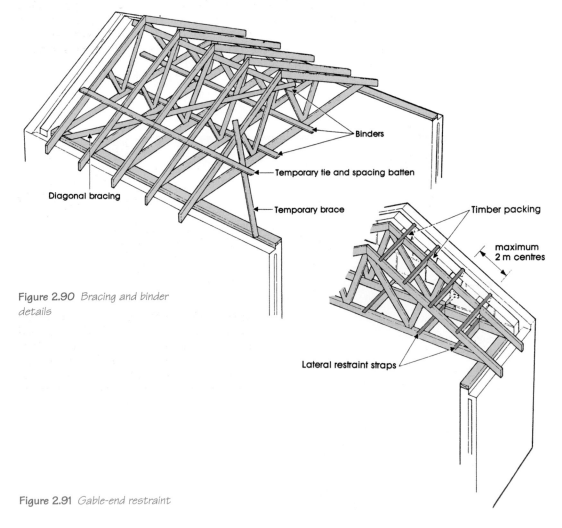

Figure 2.90 *Bracing and binder details*

Figure 2.91 *Gable-end restraint*

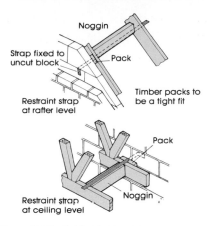

Noggin

Strap fixed to
uncut block

Pack

Restraint strap
at rafter level

Timber packs to
be a tight fit

Pack

Restraint strap
at ceiling level

Noggin

Figure 2.92 *Gable tied back to roof for support*

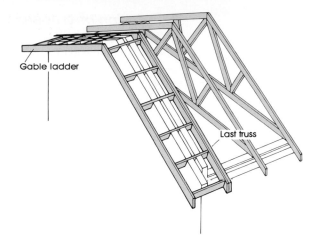

Gable ladder

Last truss

Figure 2.93 *Prefabricated gable ladder*

Prefabricated gable ladders (Figure 2.93) are fixed to the last truss when an overhanging verge is required.

Trimming

safety tip

Trussed rafters that are trimmed or altered without the structural designer's approval will seriously weaken the roof and may result in structural collapse.

Wherever possible, openings in roofs for chimney stacks and loft hatches should be accommodated within the trussed rafter spacing. If larger openings are required the method shown in Figure 2.94 can be used. This entails positioning a trussed rafter on either side of the opening and infilling the space between with normal rafters, purlins and ceiling joists. For safety reasons, on no account should trussed rafters be trimmed or otherwise modified without the structural designer's approval.

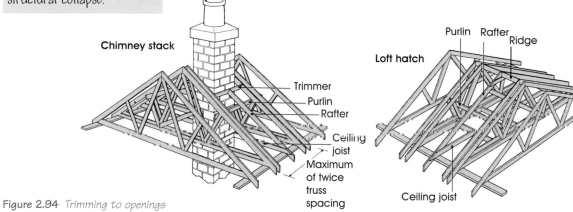

Chimney stack

Trimmer
Purlin
Rafter

Ceiling joist

Maximum of twice truss spacing

Loft hatch

Purlin   Rafter
Ridge

Ceiling joist

Figure 2.94 *Trimming to openings*

Water-tank platforms – these should be placed centrally in the roof with the load spread over at least three trussed rafters. The lower bearers of the platform should be positioned so that the load is transferred as near as possible to the mid-third points of the span (see Figure 2.95).

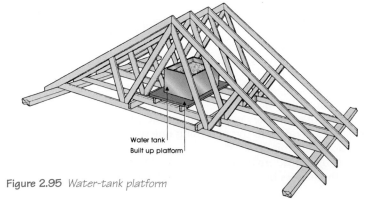

Water tank
Built up platform

Figure 2.95 *Water-tank platform*

Carcassing **Chapter 2**

Erection of trussed rafters

Trussed rafters may be lifted into position with the aid of a crane, either singly from the node joints using a spreader bar and slings, or in banded sets. In both cases these should be controlled from the ground using a guide rope to prevent swinging. In addition, where large cranes are available, the entire roof may be assembled, braced, felted and battened at ground level and then lifted into position as one whole weatherproof unit, requiring only tiling at a later stage. This method is particularly suited to the rapid erection of timber frame buildings. Figure 2.96 illustrates the handling of trussed rafters.

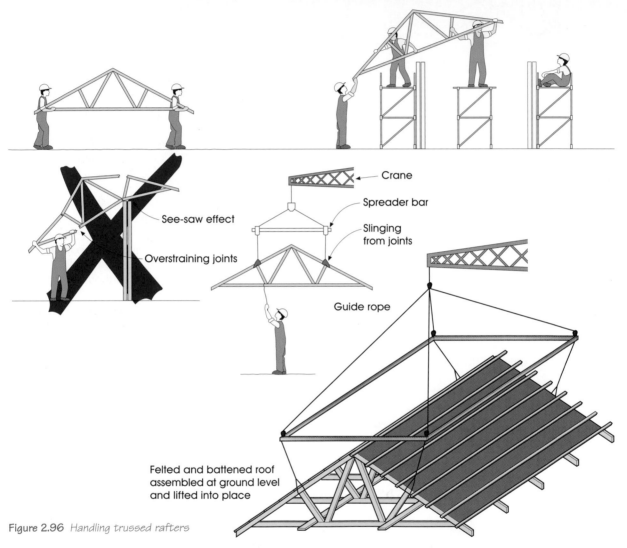

Figure 2.96 *Handling trussed rafters*

The erection procedure – the erection procedure for a gable-end roof using trussed rafters is as follows:

1. Mark the position of the trusses along the wall plates (see Figure 2.97).
2. Once up on the roof, the first trussed rafter can be placed in position at the end of the under-rafter diagonal bracing. It can then be fixed at the eaves, plumbed and temporarily braced.
3. Fix the remaining trussed rafters in position one at a time to the gable end, temporarily tying each rafter to the preceding one with a batten.
4. Fix diagonal bracing and binders.
5. Repeat the previous procedure at the other end of the roof.
6. Position and fit the trusses between the two braced ends one at a time, and fix binders.

7. Fix ladder frames and restraint straps.
8. Finish the roof at the eaves and verge with fascia, bargeboard and soffit as required.

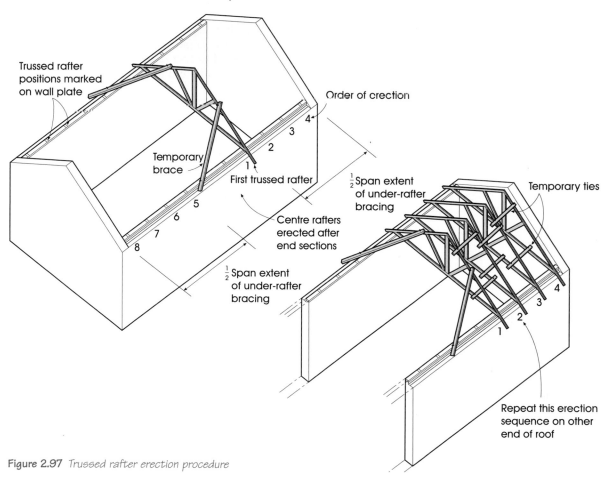

Trussed rafter positions marked on wall plate

Order of erection

Temporary brace

First trussed rafter

½ Span extent of under-rafter bracing

Temporary ties

Centre rafters erected after end sections

½ Span extent of under-rafter bracing

Repeat this erection sequence on other end of roof

Figure 2.97 *Trussed rafter erection procedure*

measuring up

51. Produce a sketch to show the difference between a single and double roof.

52. Define a flat roof.

53. State the purpose of the following:
   (a)   binder   (b)   wall plate   (c)   ridge.

54. Explain why firring pieces are used in flat roofs.

55. The joint used at the intersection of a rafter and wall plates is:
   (a)   birdsmouth   (c)   half lapped joint
   (b)   splayed dovetail   (d)   butt joint.

56. Describe the typical procedure for the erection of a gable-end cut roof.

57. Explain the reason why a purlin is incorporated into a roof.

58. Produce a sketch to show and state the purpose of lateral restraint straps used in roofing structures.

Carcassing **Chapter 2**

Verge and eaves finishings

Verge, pitched roof

To finish the verge of a gable-end roof the ridge and wall plate are extended past the gable-end wall and an additional rafter is pitched to give the required gable overhang. Noggins are fixed between the last two rafters to form a gable ladder. This provides a fixing for the bargeboard, soffit and tile battens (see Figure 2.98).

Bargeboard

A bargeboard is the continuation of a fascia board around the verge or sloping edge of the roof (typically from 25 mm × 150 mm planed-all-round – PAR – softwood). It provides a finish to the verge. The lower end of the bargeboard is usually built up to box in the wall plate and eaves. A template may be cut for the bargeboard eaves shaping in order to speed up the marking out where a number of roofs are to be cut and also to ensure that each is the same shape (especially where more elaborate designs are concerned).

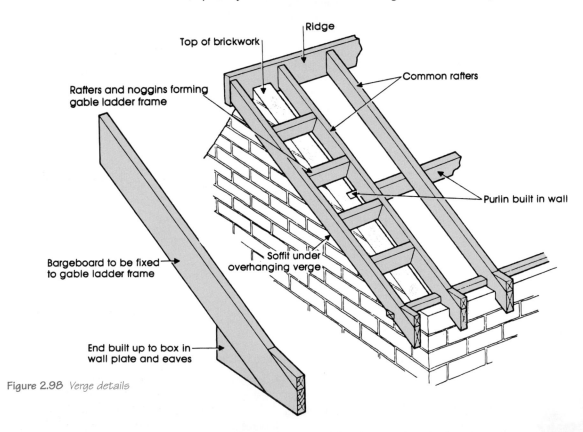

Ridge

Top of brickwork

Common rafters

Rafters and noggins forming gable ladder frame

Purlin built in wall

Bargeboard to be fixed to gable ladder frame

Soffit under overhanging verge

End built up to box in wall plate and eaves

Figure 2.98 *Verge details*

did you know?

As well as fixing the roof tiles and battens, the roof tiler will also lay a sarking felt under the tile battens, which provides protection from wind-driven rain entering the roof space.

Two methods may be used for determining the bevels at the apex (top) and foot (bottom) of the bargeboard:

- ◆ **Marking in position** – the board is temporarily fixed in position, so a spirit level can then be used to mark the plumb cut (vertical) and seat cut (horizontal) in the required positions (see Figure 2.99).
- ◆ **Determining bevels** – adjustable bevel squares may be set to the required angles for the plumb and seat cuts, using a protractor. These angles will be related to the pitch of the roof (see Figure 2.100).

The sum of all three angles in a triangle will always be 180°. Therefore in a 30° pitched roof, the apex angle will be 120° (180° – twice pitch), making the plumb cut for each bargeboard 60° (half of apex angle). The seat cut is at right angles to the plumb cut and is at the same angle as the roof pitch (see Figure 2.101).

did you know?

Bargeboards may also be marked out with the aid of a roofing square, using the same plumb and seat cut settings used for the rafter.

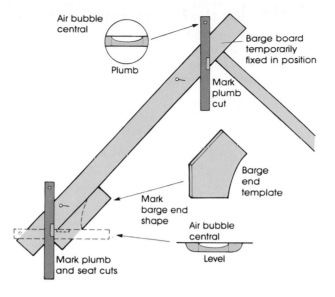

Air bubble central

Plumb

Barge board temporarily fixed in position

Mark plumb cut

Barge end template

Mark barge end shape

Air bubble central

Level

Mark plumb and seat cuts

Figure 2.99 *Marking out a bargeboard*

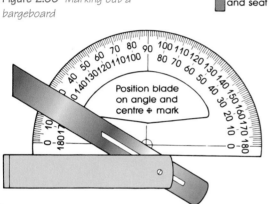

Position blade on angle and centre ⊕ mark

Figure 2.100 *Setting an adjustable bevel to a known angle*

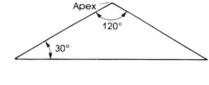

Apex

120°

30°

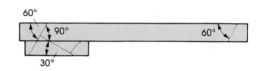

60°

90°

60°

30°

Figure 2.101 *Determining angles for a bargeboard*

activity

Determine and sketch on the protractors the adjustable bevel set up for the plumb and seat cut angles of a 40° pitched roof.

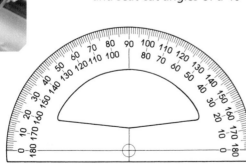
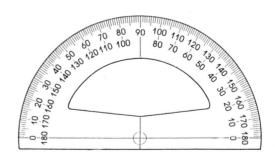

Figure 2.102 *Protractors for setting up adjustable bevels to plumb and seat cut angles*

Note: You may photocopy this activity box or download it from www.nelsonthornes.com/carpentry

Fixing the bargeboard

The foot of a bargeboard may be either mitred to the fascia board, butted and finished flush with the fascia board or butted and extended slightly in front of the fascia board (see Figure 2.103). The actual method used will depend on the specification and/or supervisor's instructions.

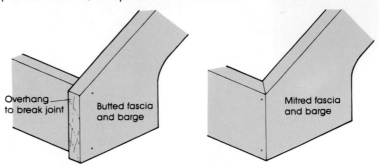

Figure 2.103 *Jointing barge to fascia board*

The mitred joint is preferred for high-quality work. The angle of the mitre for the bargeboard and fascia is best marked in position. Temporarily fix each in position, one at a time. Use a piece of timber of the same thickness to mark two lines across the edge of the board and join the opposite corners to form the mitre. The face angle will be 90° for the fascia board and a plumb cut for the bargeboard (see Figure 2.104).

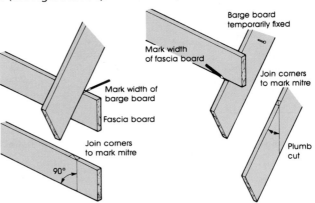

Figure 2.104 *Marking barge to fascia board mitre*

Where timber of sufficient length is not available for a continuous barge, splayed heading joints may be used, as shown in Figure 2.105.

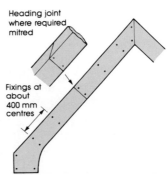

Figure 2.105 *Bargeboard lengthening/fixings*

After marking, cutting to shape, mitring and fitting, the bargeboard can be fixed to its gable ladder by double nailing at approximately 400 mm centres. Use either oval nails, wire nails, lost-head nails or cut nails. These should be at least two-and-a-half times the thickness of the bargeboard in length in order to provide a sufficiently strong fixing. For example, an 18 mm thick bargeboard would require nails at least 45 mm long (50 mm being the nearest standard length). All nails should be punched below the surface ready for subsequent filling by the painter.

Eaves, pitched roof

These may be finished (see Figure 2.106) as either:

◆ flush;
◆ overhanging, open or closed; or
◆ sprocketed.

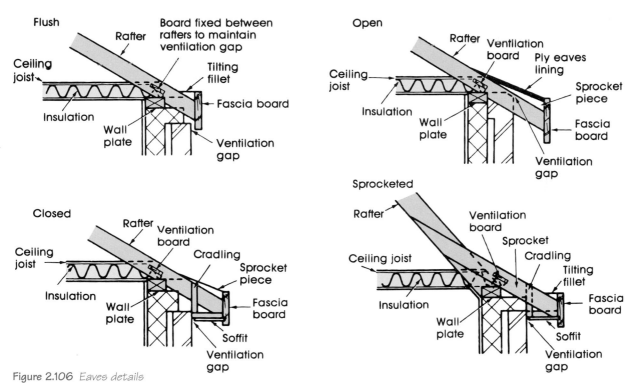

Figure 2.106 *Eaves details*

Flush eaves

In this method the ends of the rafters are cut off 10–15mm past the face of the brickwork and the fascia board is nailed directly to them to provide a fixing point for the gutter. The small gap between the back of the fascia board and the brickwork allows for roof space ventilation.

Open eaves

Open eaves project well past the face of the wall to provide additional weather protection. The ends of the rafters should be prepared as they are exposed to view from the ground. In cheaper quality work the fascia boards are often omitted and the gutter brackets fixed directly to the side of the rafter.

Closed eaves

These overhang the face of the wall in the same way as open eaves, except that the ends of the rafters are closed with a soffit. Cradling brackets are nailed to the sides of the rafters to support the soffit at the wall edge.

Sprocket piece

Flush, open and closed eaves often use sprocket pieces nailed to the top of each rafter to reduce the pitch of the roof at the eaves. This has the effect of easing fast-flowing rainwater under storm conditions into the gutter.

Sprocketed eaves

On steeply pitched roofs the flow of rainwater off the roof surface has a tendency to overshoot the gutter. Sprockets can be nailed to the side of each

did you know?

Where oval or cut nails are used they should be positioned so that their greatest width runs with the timber's grain, which reduces the likelihood of splitting the grain.

rafter to lower the pitch and slow down the rainwater before it reaches the eaves. This reduces the likelihood of rainwater overshooting the eaves and/or hitting the front of the gutter and splashing back soaking the eaves timbers, with the subsequent risk of rot. In addition the use of sprockets also enhances the appearance of a roof giving it a distinctive 'bell-cast' appearance.

Boards and finishes

Fascia board

The fascia board is the horizontal board (typically ex 25 mm × 150 mm PAR softwood), which is fixed to the ends of the rafters, to provide a finish to the eaves and a fixing for the guttering.

Before fixing the fascia board the rafter feet will require marking and cutting to plumb and line as shown in Figure 2.107; a seat cut may also be required depending on the assembly detail.

◆ Measure out from brickwork the required soffit width and mark on the last rafter at either end of the roof.
◆ Mark the plumb cut and the seat cut if required using a spirit level.
◆ Stretch a string line between the end two rafters and over the tops of the other rafters; use a spirit level to mark each individual plumb cut.
◆ Cut the plumb cuts using either a hand saw or portable circular saw.
◆ Where a seat cut is required, move the line down to the seat cut position on the end rafters; use a spirit level to mark each individual seat cut.
◆ Cut the seat cuts using either a hand or portable circular saw.

Where timber of sufficient length is not available for a continuous fascia board, splayed heading joints may be used. Figure 2.108 shows how these should be positioned centrally over a rafter end.

Where level fascia boards are returned around corners it is standard practice to use a mitre at the external and butt at the internal corners. Both of these joints should be secured by nailing (50 mm ovals).

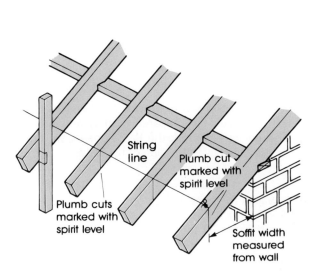

Figure 2.107 *Marking out plumb cut at eaves*

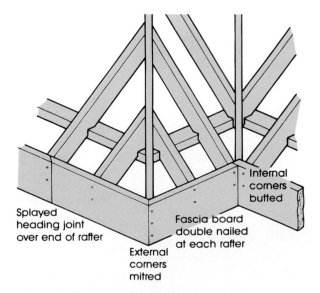

Figure 2.108 *Jointing of fascia board*

Fascia boards are fixed to the end of each rafter using two nails. These are normally oval nails, wire nails, lost-head nails or cut nails, at least two-and-a-half times the thickness of the fascia board in length. Typically 50 mm or 62 mm nails provide a sufficiently secure fixing. All nails should be punched below the surface ready for subsequent filling by the painter.

Prior to final fixing the fascia should be checked for line as shown in Figure 2.109.

◆ Drive nails on the face of the fascia at each end of the roof. Strain a line between them.
◆ Cut three identical pieces of packing, place one at each end under the line and use the third to check the distance between the fascia and the line at each rafter position.
◆ Pack out or use a saw to ease ends of rafters as appropriate, so that the packing piece just fits between the fascia and the line.

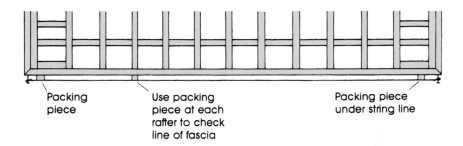

Packing piece

Use packing piece at each rafter to check line of fascia

Packing piece under string line

Figure 2.109 Checking fascia board for line

Soffit board

Verge, eaves and soffit finishings in PVCu – increasingly extruded PVCu (unplasticized polyvinyl-chloride) sections are being used for new-build and refurbishment works. They have the advantage over timber in that they require very little in the way of ongoing maintenance, whereas timber can suffer from the effects of weather and requires regular re-painting.

Standard profiles are available for fascias, soffits and bargeboards as illustrated in Figure 2.110. Corner and butt joint trims are also available, as are matching soffit ventilation sections. Profiles are easily cut to length using a fine tooth panel or dovetail saw. Fixing is normally directly into the rafter or cradling at about 600 mm centres, using stainless steel white plastic dome-headed pins.

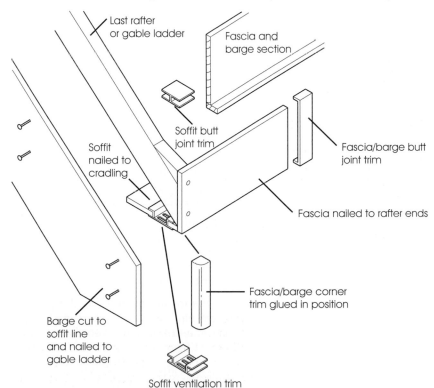

Last rafter or gable ladder

Fascia and barge section

Soffit butt joint trim

Soffit nailed to cradling

Fascia/barge butt joint trim

Fascia nailed to rafter ends

Fascia/barge corner trim glued in position

Barge cut to soffit line and nailed to gable ladder

Soffit ventilation trim

Figure 2.110 Use of PVCu profiles for eaves and verge finishing

Preservative treatment

It is recommended that all timber used for verge and eaves finishes is preservative treated before use. Any preservative-treated timber cut to size on site will require re-treatment on the freshly cut edges/ends. This can be carried out by applying two brush flood coats of preservative.

Thermal insulation

Thermal insulation was normally done by placing insulation between the ceiling joists and incorporating a vapour check, such as foil-backed plasterboard, at ceiling level. However with the increased levels of thermal insulation now required for new buildings, a second layer of insulation is often laid over the ceiling joists (Figure 2.111).

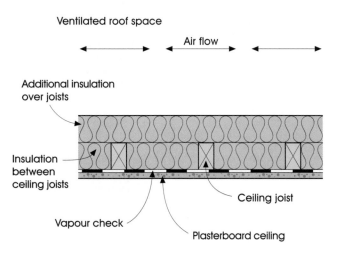

Figure 2.111 *Typical pitch roof insulation between and over ceiling joists*

Roof ventilation

Care must be taken not to block the eaves with the insulating material, as the roof space must be well ventilated in order to prevent any warm, moist air that passes through the ceiling from condensing in the roof space.

To reduce the likelihood of condensation within the roof space ventilation is required by the Building Regulations. All roofs must be cross-ventilated at eaves level by permanent vents. These must have an equivalent area equal to a continuous gap along both sides of the roof of 10 mm or 25 mm where the pitch of the roof is less than 15°. (See Figure 2.112.)

This ventilation requirement can be achieved by:

◆ leaving a gap between the wall and soffit (this may be covered with a wire mesh to prevent access by birds, rodents and insects etc.);
◆ using a proprietary ventilation strip fixed to the back of the fascia;
◆ using proprietary circular soffit ventilators let into the soffit at about 400 mm centres.

Lean-to roofs must have high-level ventilation in the roof slope close to the abutting wall. Proprietary vents are available for this purpose; these must have an equivalent area equal to a continuous gap along the abutment of 5 mm.

Where the insulation follows the pitch of the roof, there must be at least a 50-mm air space between the insulation and the underside of the roof structure; and high-level ventilation at the ridge equal to a continuous gap of at least 5 mm. Proprietary ridge vents are available for this purpose.

Roofs where the span exceeds 10 m or non-rectangular roofs may require additional ventilation, totalling 0.6% of the roof area.

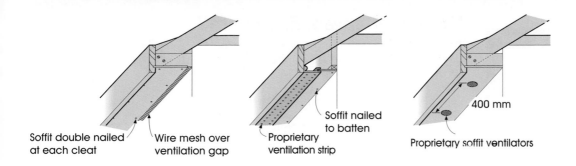

Soffit double nailed at each cleat

Wire mesh over ventilation gap

Proprietary ventilation strip

Soffit nailed to batten

Proprietary soffit ventilators

400 mm

Lean-to roof

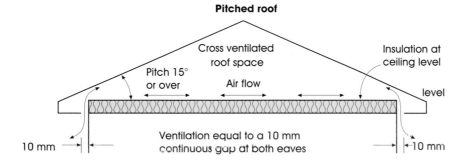

High level ventilation equal to a 5 mm continuous gap

Ventilated roof space

Insulation at ceiling level

Air flow

Pitch 15° or over

Eaves ventilation equal to a 10 mm continuous gap

10 mm

Pitched roof

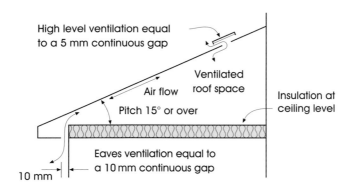

Cross ventilated roof space

Insulation at ceiling level

Pitch 15° or over

Air flow

level

10 mm

Ventilation equal to a 10 mm continuous gap at both eaves

10 mm

Attic pitched roof

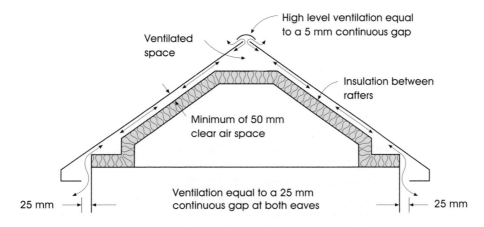

High level ventilation equal to a 5 mm continuous gap

Ventilated space

Insulation between rafters

Minimum of 50 mm clear air space

25 mm

Ventilation equal to a 25 mm continuous gap at both eaves

25 mm

Figure 2.112 *Pitched roof ventilation*

did you know?

Capillary attraction or capillarity is the phenomenon whereby a liquid can travel against the force of gravity, even vertically in fine spaces or between surfaces placed closely together. This is due to the liquid's own surface tension: the smaller the space the greater the attraction. Measures taken to prevent capillarity, such as forming a drip or groove, are known as anti-capillary measures.

Verge and eaves, flat roofs

The finishing of verge and eaves to flat roofs is a similar process to that of pitched roofs. Both can be finished as either flush or overhanging details as shown in Figures 2.113 and 2.114.

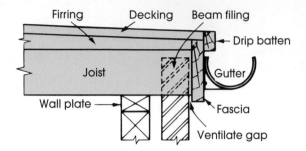

Figure 2.113 *Flush eaves*

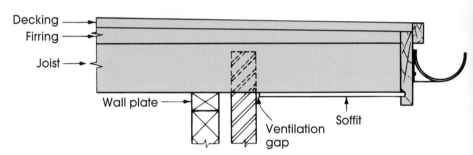

Figure 2.114 *Overhanging eaves*

A drip batten is fixed at the lower eaves to the top of the fascia to extend the roof edge into the gutter (see Figure 2.115). This extension enables the roofing felt to be turned around it. Rainwater flows off the drip batten into the centre of the gutter, thus ensuring efficient discharge and reducing the risk of rot damage to timber. A drip batten is also fixed at higher eaves and verge edges, again to enable roofing felt to be turned around it. Any moisture is then allowed to drip clear of the fascia and not creep back under the felt by capillary attraction, with the subsequent likelihood of rot.

Due to the depth of joists deep fascia boards are often required. These may be formed from solid timber, plywood (WBP, weather and boil proof) or matchboarding. An alternative is to reduce the depth of the joists at the ends by either a splay or square cut as shown in Figure 2.116.

Figure 2.117 shows how a boxed or internal gutter can be formed at the eaves as an alternative to external gutters. The gutter fall (1:60) can be achieved by progressively increasing the depth of cut-out in the joist ends towards the outlet.

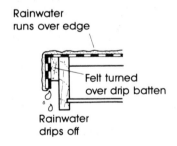

Figure 2.115 *Use of drip batten to eaves and verge of flat roof*

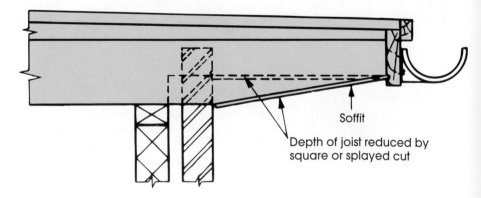

Figure 2.116 *Reducing the depth of the joist*

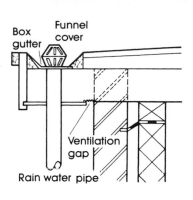

Figure 2.117 *Boxed gutter to a flat roof*

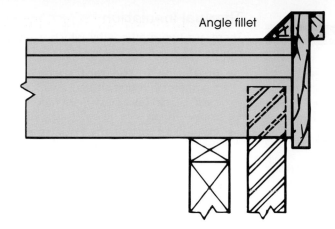

Figure 2.118 *Angle fillet at eaves*

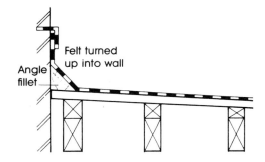

Figure 2.119 *Use of an angle fillet to abutments on flat roofs*

Angle fillets, as shown in Figures 2.118 and 2.119, are fixed around the upper eaves, verge and edges of the roof, which abut brickwork. They enable the roofing felt to be gently turned at the junction; sharp turns on felt lead to cracking, subsequent leaking and risk of rot. In addition, the use of angle fillets also prevents rainwater from dripping or being blown over the edges of the roof.

Overhanging details for the verge and eaves positioned at right angles to the main joist run can be formed using two alternative methods:

◆ *Returned eaves* – short joists are fixed at right angles to the last main joist using joist hangers (see Figure 2.120).
◆ *Ladder frame eaves* – a ladder frame made from two joists and noggins is made up and fixed to the last joist (see Figure 2.121).

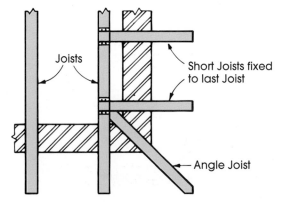

Figure 2.120 *Returned eaves*

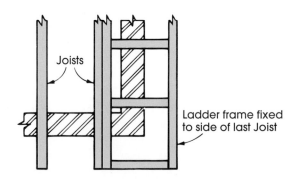

Figure 2.121 *Ladder frame eaves*

Carcassing **Chapter 2**

Thermal insulation

There are two methods by which a flat roof can be insulated, as shown in Figure 2.122.

◆ **Insulation in the roof space** – known as cold roof construction, this has thermal insulation and a vapour check at ceiling level. The roof space itself is cold and must be vented to the outside air to prevent interstitial condensation.

◆ **Insulation above the roof space** – known as warm roof construction, this has its thermal insulation and vapour barrier placed over the roof decking. The roof space is kept warmer than the outside air temperature and does not require ventilation.

The actual thickness of insulation used will be dependent on its type, positioning and the use of the building. The Building Regulations should be consulted for further information.

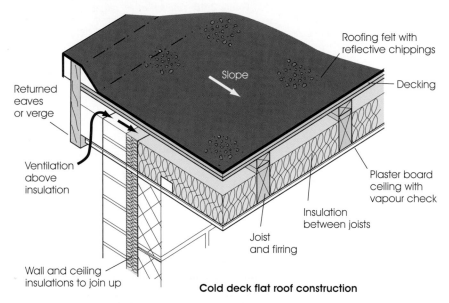

Cold deck flat roof construction

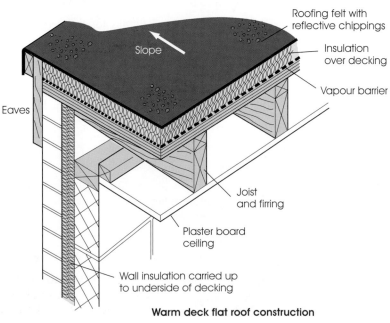

Warm deck flat roof construction

Figure 2.122 *Flat roof thermal insulation*

Ventilation

To reduce the likelihood of condensation within the roof space, ventilation is required by the Building Regulations. All cold flat roofs must be cross-ventilated at verge or eaves level by permanent vents. These must have an equivalent area equal to a continuous gap along both sides of the roof of 25 mm. The void in the roof between the insulation and the underside of the decking should be a minimum of 50 mm to allow free air movement (see Figure 2.123). Joists that run at right angles to the airflow should be counter-battened before decking to provide the free air space.

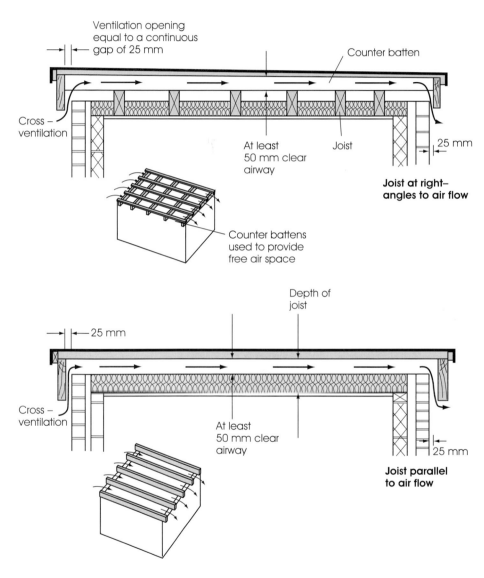

Figure 2.123 *Flat roof ventilation*

Either of the following can achieve this ventilation requirement:

◆ leaving a gap between the wall and soffit (this may be covered with a wire mesh to prevent access by birds, rodents and insects, etc.);
◆ the use of proprietary ventilation systems in the soffit.

Where cross-ventilation is not possible due to abutting walls a warm roof construction should be used.

Determining materials

Calculating the lengths of materials required for fascia boards, bargeboards and soffits is often a simple matter of measuring, allowing a certain amount extra for jointing, and adding lengths together to determine total metres run.

example

A hipped-end roof requires two 4.4 m lengths and two 7.2 m lengths of ex 25 mm × 150 mm PAR softwood for its fascia boards.

Metres run required = (4.4 x 2) + (7.2 x 2)

= 8.8 + 14.4 = 23.2 m

It is standard practice to allow a certain amount extra for cutting and jointing. This is often 10%. Ten per cent of any number can be found by moving its decimal point one place forward.

10% of 23.2 = 2.32

The total metres run of timber is determined by adding the percentage increase to the original number.

Total metres run required = 23.2 + 2.32 = 25.52 m

example

The length of timber needed for bargeboards may require calculation using Pythagoras' theorem of right-angled triangles. This states that in any right-angled triangle the square of the hypotenuse (longest side) is equal to the sum of the square of the other two sides.

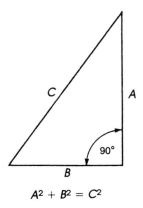

$$A^2 + B^2 = C^2$$

Figure 2.124 *Pythagoras' theorem*

Thus the length of the hypotenuse in a triangle having sides A, B, C is:

$C = \sqrt{(B^2 + A^2)}$

Determine the length of bargeboard required for one gable having a rise of 3 m and a span of 7 m.

A = Half span = 3.5 m

B = Rise = 3 m

C = Length of barge required for one slope

= $\sqrt{(B^2 + A^2)}$

= $\sqrt{[(3 \times 3) + (3.5 \times 3.5)]} = \sqrt{(9 + 12.25)}$

= $\sqrt{21.25}$ = 4.610 m

Total length of bargeboard required for one gable end is 9.220 m

Where sheet material is used for fascias and soffits the amount that can be cut from a full sheet often needs calculating. This entails dividing the width of the sheet by the width of the fascia or soffit; then using the resulting whole number to multiply by the sheet's length, to give the total metres run.

example

Determine the total metres run of 150mm wide soffit board that may be cut from a 1220mm × 2440mm sheet.

$$\text{Number of lengths} = 1220 \div 150$$
$$= 8.133, \text{ say} = 8$$
$$\text{Total metres run} = 8 \times 2.440 = 19.520\,\text{m}$$

activity

A 6.5m × 2.8m plan gable-end roof has a rise of 1.4m and an eaves and verge overhang of 175mm. Determine the total amount of ex 25mm × 225mm PAR softwood required for the fascia and bargeboards and the number of whole 1220mm × 2440mm sheets required to cut the 150-mm wide eaves and verge soffit boards. Allow 10% to the total amount as a cutting allowance.

If you are unfamiliar with calculations or simply want to 'brush up' before attempting this learning task, refer to *A Building Craft Foundation*, 3rd edition, Chapter 5: Numerical Skills.

measuring up

59. Name the regulations that apply to the provision of ventilation in roof spaces.

60. State the purpose of soffit ventilators.

61. Produce sketches to identify the following:
    (a) flush eaves
    (b) open eaves
    (c) closed eaves.

62. State the purpose of a heading joint and produce a sketch of its use in a fascia board.

63. State the purpose of sprockets used in pitched roofs.

64. Produce a sketch to illustrate a sprocket.

65. State the purpose of angle fillets and drip battens used in flat roofs.

66. Produce a sketch to illustrate angle fillets and drip battens.

67. State the reason why the ends of timber sawn on-site should be treated with a preservative.

68. Describe the application of preservative on-site to a freshly cut end.

Carcassing **Chapter 2**

69. Describe each of the following:
    (a)   fascia board
    (b)   bargeboard
    (c)   soffit board.

70. Name a nail suitable for fixing a 6 mm non-combustible sheet material soffit to softwood cradling.

71. Name two nails suitable for fixing an ex 25 mm × 150 mm softwood fascia board to the rafter ends.

72. State the advantage that PVCu has over timber when used for fascia and bargeboards.

73. Produce a sketch to illustrate the use of PVCu profiles to finish the eaves of a pitched roof.

74. Produce a sketch to illustrate a method of determining the plumb cut required for the upper end of a bargeboard.

First Fixing

This chapter is intended to provide the reader with an overview of first fixing work. Its contents are assessed in the NVQ Unit No. VR 09 Install First Fixing Components.

In this chapter you will cover the following range of topics:

◆ Timber studwork partitions
◆ Straight flight stairs
◆ Frames and linings.

What's required in VR 09 Install First Fixing Components?

To successfully complete this unit you will be required to demonstrate your skill and knowledge of the following first fixing elements:

◆ partition walls;
◆ straight flight stairs;
◆ frames and linings.

You will be required practically to:

◆ install door and window frames;
◆ install door and or hatch linings;
◆ lay flat roof decking or flooring;
◆ install timber partitions, including fixing plasterboard;
◆ install staircases.

Note: The laying of flat roof decking or flooring is also a requirement of NVQ Unit No. VR 11 Erect Structural Carcassing Components. Once suitable evidence has been obtained it can be used for both units. Except for partitions and staircases, evidence for the successful completion of all other practical tasks must be work based.

Timber studwork partitions

did you know?

First fixing is the work undertaken by the carpenter after carcassing but before the building is plastered internally. Plumbers and electricians also use the term for the running of service pipes and cables.

Partition terminology

Partition – an internal wall used to divide space into a number of individual areas or rooms. Partitions are normally of a non-structural/non-load-bearing nature. Figure 3.1 shows how they are commonly formed from timber studwork framing, proprietary systems or lightweight blockwork.

Studwork framing/partition – commonly known as stud partitioning. This is a partition wall built of timber or metal studs, fixed between a sole and head plate, often incorporating noggins for stiffening and fixing (see Figure 3.2).

Stud – a vertical timber or metal member of a partition wall fixed between the sole plate and head plate. The main member of a partition, it provides a fixing for the covering material.

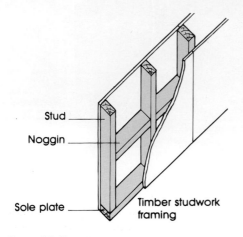
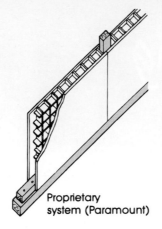
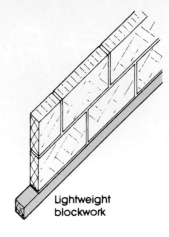

Figure 3.1 *Partitions*

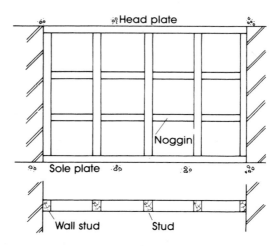

Figure 3.2 *Stud partition components*

did you know?

Timber stud partitions in modern buildings are normally not load bearing and are used just to divide up space. In older buildings stud partitions may transfer imposed loads from the floor or roof above.

Sole plate/head plate – a horizontal timber fixed above or below studs to provide a fixing point for the studs and ensure an even distribution of loads.

Noggin – a short horizontal piece of timber fixed between vertical studs of a partition. Its use stiffens the studs, provides an intermediate fixing point for the covering material and in addition a fixing point for heavy items that may be hung on the partition (WC cistern, hand basin etc.).

Timber stud partitions

Traditionally, timber is used for making studwork framing. Planed-all-round (PAR) timber is preferred not only because of its uniform cross-section but also because it is better to handle. Machined timber, also termed regularized (timber machined to a consistent width by re-sawing or planing one or both edges) or ALS/CLS (American/Canadian Lumber Stock) processed timber with rounded corners, may also be specified.

A consistent-sized section aids the plumbing of a partition and also provides a flat fixing surface for the partition covering materials (see Figure 3.3).

Covering materials

The standard covering material for partition walls is plasterboard. Typical for stud partition use are thicknesses of 9.5 mm or 12.5 mm sheet sizes of either 900 mm × 1800 mm or 1200 mm × 2400 mm. Tapered edge boards are best for a dry finish. These should be filled and taped to provide a smooth seamless joint that will not show through the subsequent paint or wallpaper finish.

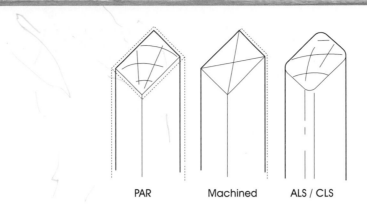

Figure 3.3 *Consistent-size timber sections*

PAR   Machined   ALS / CLS

Square-edge boards can also be used for a filled and taped dry finish, although they are primarily used for surfaces that are skimmed with a thin coat of wet board finish plaster (see Figure 3.4).

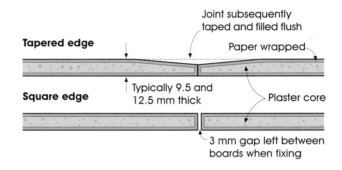

Figure 3.4 *Types of plasterboard*

Jointing partition members

Traditionally, timber partitions were framed using basic joints, to locate members and provide strength, as shown in Figure 3.5.

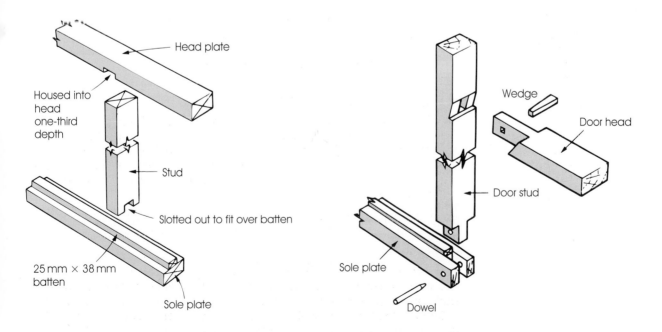

Figure 3.5 *Traditional joint details*

Studs were often housed into the head plate, slotted over a batten at the sole plate and mortised or tenoned at openings. However, present-day techniques, calling for speed of erection and economy result in the majority of partitions being simply butt jointed and skew nailed (Figure 3.6). An alternative to both methods would be the use of metal framing anchors. These provide a quick yet strong fixing, although they are rarely specified.

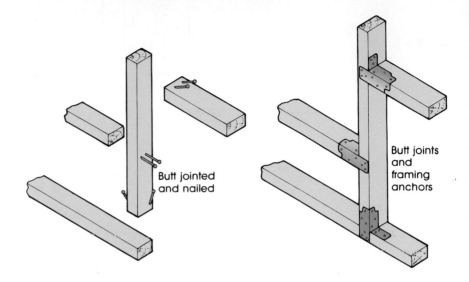

Figure 3.6 *Alternative joint details*

Constructing in situ stud partitions

Mark the intended position of the partition on the ceiling. Ideally this should either be at right angles to the joists or positioned under a joist or double joist. Where the joists run in the same direction as the partition and it is not directly under a joist, noggins will have to be fixed between joists, to provide a fixing point (see Figure 3.7).

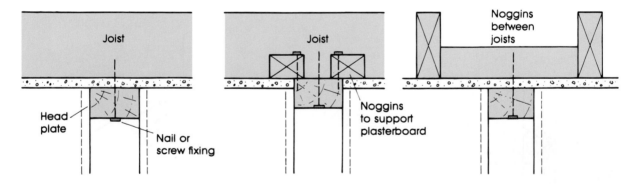

Figure 3.7 *Fixing of head plate*

Fix the head plate to ceiling, using 100 mm wire nails or oval nails at each joist position or 400 mm centres as appropriate. In older buildings, where ceilings might be in a poor position or easily damaged by nailing, 100 mm screws can be used as an alternative.

Plumb down from the head plate on one side at each end, to the floor, using either a straight edge and level or a plumb bob and line as shown in Figure 3.8. This establishes the position of the sole plate.

Fix the sole plate in position with 100 mm wire nails, oval nails or screws as appropriate. Ideally, as with head plates, this should be either at right angles to

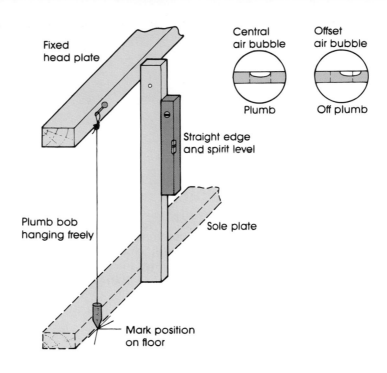

Figure 3.8 *Plumbing partition*

the joists or over a joist or double joist. Where the joists run in the same direction as the partition but not directly under it, noggins will have to be fixed between joists to provide a fixing point (see Figure 3.9). Where the sole plate is fixed to a concrete floor this should be plugged and screwed.

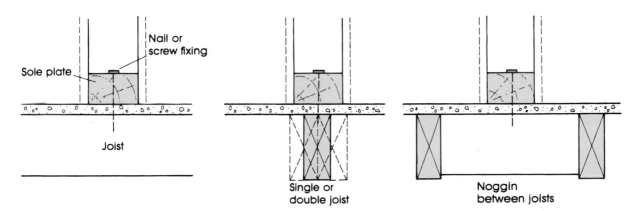

Figure 3.9 *Fixing of sole plate*

It may be possible to 'shot fix' both head and sole plates to concrete surfaces. However, this is not within the scope of this package. Please refer to your supervisor for permission/instruction.

Cut, position and fix the end wall studs (see Figure 3.10). These may be either:

◆ plugged and screwed to brickwork and blockwork using 100 mm screws into proprietary plugs or twisted timber pallets;
◆ nailed to blockwork or into brickwork mortar joints using 100 mm cut nails;
◆ nailed directly to brickwork using 75 mm hardened steel masonry nails.

safety tip

It is essential that eye protection is used when driving masonry nails as they are liable to shatter.

First Fixing

Chapter 3

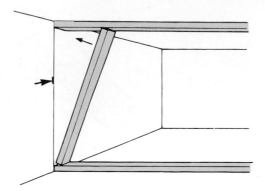

Figure 3.10 *Positioning wall stud*

Mark the positions of vertical studs on the sole and head plates. Studs will be required at each joint in the covering material and using 1200 mm wide sheets at 400 mm centres for 9.5 mm plasterboard or at 600 mm centres for 12.5 mm plasterboard; using 900 mm wide sheets, all studs should be at 450 mm centres (see Figure 3.11).

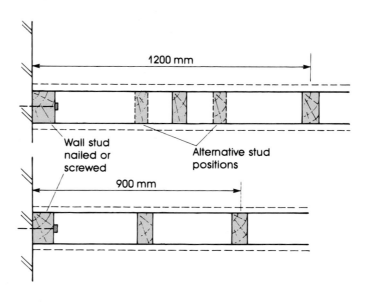

Figure 3.11 *Stud spacings*

Similar centres may be employed for other covering material/cladding. Assuming 12.5 mm × 1200 mm covering, the second stud is fixed with its centre 600 mm from the wall, the third with its centre 1200 mm from the wall and the remaining studs at 600 mm centres thereafter.

Measure, cut and fix each stud to head and sole plates, using 100 mm wire or oval nails, driven at an angle (skew nailed, see Figure 3.12). Each stud should be measured and cut individually as the distance between the plates may vary along their length. Studs should be a tight fit: position one end, angle the stud and drive the other end until plumb. The length of each stud may be measured using either a tape or pinch rods as shown in Figure 3.13.

Mark noggin centre line positions; these will vary depending on the specification. Typically they are fixed at vulnerable positions where extra strength is required: at knee height 600 mm up from floor; at waist height 1200 mm up from floor; and at shoulder height 1800 mm up from floor. Where deep section skirting is to be fixed a noggin may be specified near its top edge for fixing purposes as shown in Figure 3.14. Additional noggins will also be required where heavy items are to be hung on the wall and where the covering sheet material is jointed in the height of the partition. See Figure 3.15.

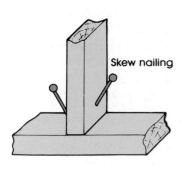

Figure 3.12 *Skew nailing*

Skew nailing

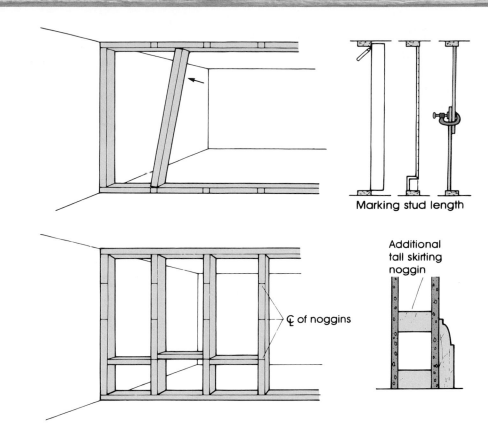

Figure 3.13 *Marking and positioning studs*

Marking stud length

Figure 3.14 *Noggin positions*

℄ of noggins

Additional tall skirting noggin

Figure 3.16 shows how to fix the noggins, either by skew nailing using 100 mm wire or oval nails or staggering either side of the centre line and through nailing using 100 mm wire or oval nails.

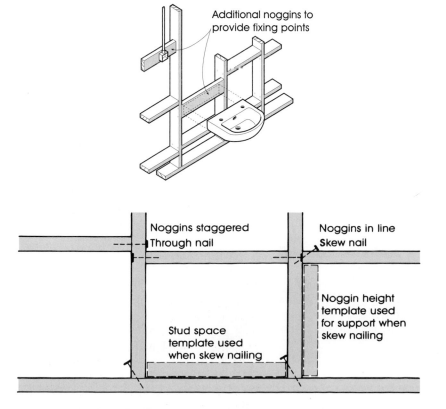

Additional noggins to provide fixing points

Figure 3.15 *Additional noggins*

Noggins staggered Through nail

Noggins in line Skew nail

Noggin height template used for support when skew nailing

Stud space template used when skew nailing

Figure 3.16 *Fixing noggins*

measuring up

1. State the purpose of a partition wall and name TWO methods of construction.

2. List FIVE component members of timber studwork framing.

3. State the difference between in situ and pre-made timber stud partitions.

4. State the purpose of noggins used in timber studwork.

5. State the purpose of noggins used between floor joists either above or below timber studwork.

6. Name a nail suitable for fixing a head plate to a timber floor joist.

7. State suitable centres for studs when using 9.5 mm thick × 1200 mm wide plasterboard covering.

8. Name a nail suitable for securing the butt joint between a stud and noggin.

9. Name a nail suitable for securing wall studs directly to brickwork and state any precautions that should be observed in its use.

10. Produce a sketch to illustrate skew nailing.

Estimating materials

To determine the number of studs required for a particular partition the following procedure, shown in Figure 3.17, can be used:

◆ Measure the distance between the adjacent walls of the room or area that the partition is to divide, say 3400 mm.
◆ Divide the distance between the walls by the specified stud spacing, say 600 mm. This gives the number of spaces between the studs. Where a whole number is not achieved round up to the whole number above. There will always be one more stud than the number of spaces, so add one to this figure to determine the number of studs. Stud centres must be maintained to suit sheet material sizes leaving an undersized space between the last two studs.
◆ The lengths of head and sole plates are simply the distance between the two walls. Each line of noggins will require a length of timber equal to the distance between the walls.

did you know?

A window in an internal wall or partition is termed a 'borrowed light' as it is said to borrow daylight from an adjacent room with external windows.

example

The total length of timber required for a partition can be determined by the following method:

7 studs at 2.4 m, 7 × 2.4	= 16.8 m
Head and sole plates at 3.4 m, 2 × 3.4	= 6.8 m
3 lines of noggins at 3.4 m, 3 × 3.4	= 10.2 m
Total metres run required, 16.8 + 6.8 + 10.2	= 33.8 m

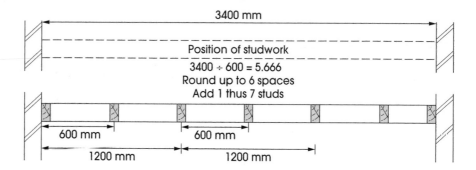

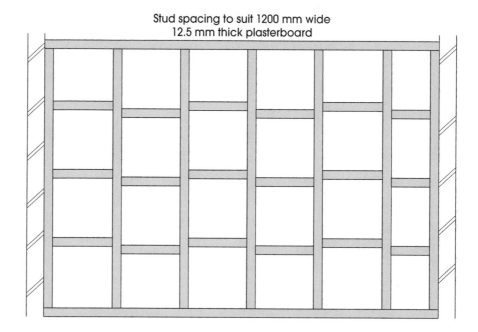

Figure 3.17 *Estimating materials for a stud partition*

Openings

Where door, serving hatch or borrowed light (internal glazing) openings are required in a studwork framing, studs and noggins should be positioned on each side to form the opening (Figures 3.18 to 3.21). The sole plate will require cutting out between the door studs. Linings are fixed around the framed opening to provide a finish and provision for hanging doors or glazing.

Returns

Should a return partition forming internal or external angles be required, stud positions must be arranged to suit. Consideration must be given to providing a support and fixing for the covering material around the return intersection. Two alternative methods, shown in Figures 3.22 and 3.23, are commonly employed. The particular method adopted will depend on whether the carpenter is fixing the covering material as the work proceeds or the plasterer is fixing it after the carpentry work is complete.

Wall plates in long partitions and those containing return intersections will require joining; preferably using halving joints secured with screws (see Figure 3.24).

Cutting and fixing plasterboard

Plasterboard is normally marked by its manufacturer to indicate which face is suitable for plastering or dry decoration. Typically the following is marked on the

First Fixing

Chapter 3

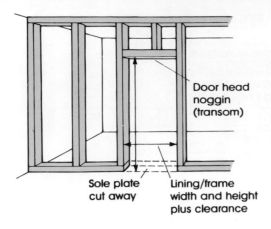

Figure 3.18 *Door opening in a stud partition*

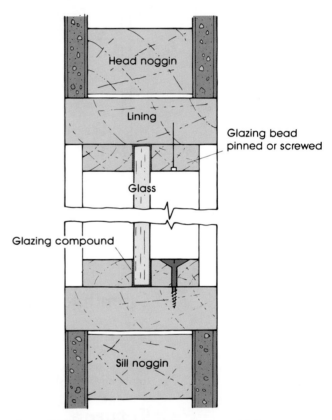

Figure 3.19 *Typical horizontal section of door opening*

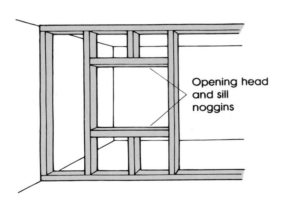

Figure 3.20 *Borrowed light opening in a stud partition*

Figure 3.21 *Typical vertical section of borrowed light*

rear face: 'Use other face for plastering and decoration.' In general the ivory face is fixed to the outside.

Cutting

Plasterboard can be cut with a saw or with a replaceable blade craft knife (see Figure 3.25).

◆ When sawing use the saw at a low angle to the sheet.
◆ When using a knife, score a fairly deep line into the face along a straight edge. Apply downward pressure to snap the sheet along the scored line. Run the knife along the snapped line on the underside to cut the paper and separate the two pieces.
◆ Holes for electrical fittings may be cut using a pad saw or with a craft knife.

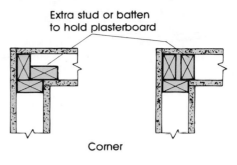

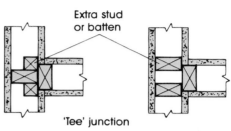

Corner

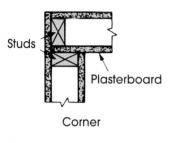

Studs

Plasterboard

Corner

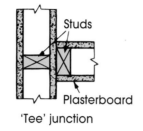

Studs

Plasterboard

'Tee' junction

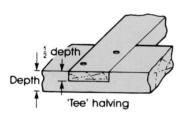

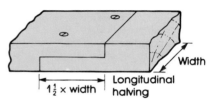

'Tee' junction

Figure 3.22 *Corner details*

Figure 3.23 *Stud treatment to provide support for coverings at return*

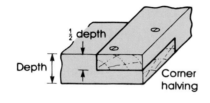

'Tee' halving

Corner halving

Longitudinal halving

Figure 3.24 *Sole and head plate joints*

Fixing procedure

Plasterboard sheets are secured to the studs using plasterboard nails or plasterboard screws at approximately 150 mm centres around the sheet edges and intermediate studs. To avoid distortion, nailing should commence from the centre of the sheet working outwards. Nails should be driven just below the surface, but taking care not to break or damage the paper face of the sheet. See Figure 3.26 for the fixing procedure:

1. Pre-cut the sheets to length, making them 10 to 15 mm shorter than the floor-to-ceiling height.
2. Starting either from the corner or a doorway opening, use a board lifter to position and hold the board in place up against the ceiling; secure to all studwork members.
3. A narrow strip must be cut from the edge of the board over a doorway, to enable two boards to butt on the centre of the stud. Alternatively a short stud can be fixed to the side of the door stud to provide a fixing.
4. Position and secure the remaining full boards in place, working across the position. Tapered edge boards can be lightly butted together; square edge boards should be fixed leaving about a 3 mm gap between them.
5. Cut final board to width, position and secure.
6. Repeat the procedure to cover the other side of the partition.

Where services are to be accommodated in the partition, they should be installed either before any plaster boarding is done or before the second face is boarded.

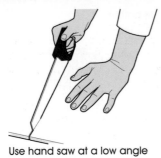

Use hand saw at a low angle

Score line on face using knife against a straight edge

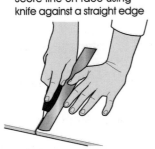

Snap board on scored line, use knife to separate

Cut holes using a pad saw

Figure 3.25
Cutting plasterboard

Pre-made partitions

Pre-made partitions may be made either by the joiner in a workshop or by the carpenter on-site. The method of jointing will normally be either studs housed into plates with noggins butt-jointed and nailed or all joints butted and nailed. In addition, occasionally pre-made partitions may be jointed using framing anchors.

This type of partition must be made under-size in both height and width, in order for it to be placed in position. Once in position folding wedges are used to take up the positioning tolerance prior to fixing (see Figures 3.27 and 3.28). Plates and wall studs are fixed using the same methods and centres as are applicable to in situ partitions.

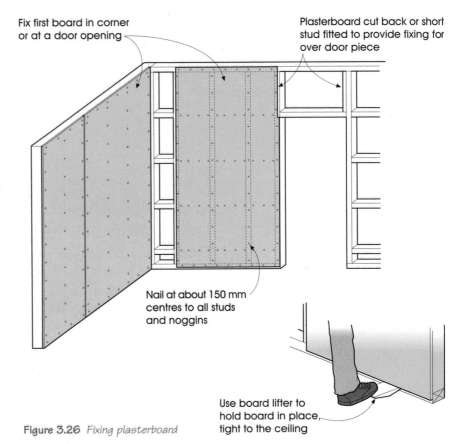

Fix first board in corner or at a door opening

Plasterboard cut back or short stud fitted to provide fixing for over door piece

Nail at about 150 mm centres to all studs and noggins

Use board lifter to hold board in place, tight to the ceiling

Figure 3.26 *Fixing plasterboard*

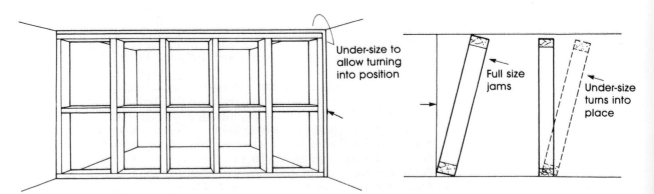

Under-size to allow turning into position

Full size jams

Under-size turns into place

Figure 3.27 *Pre-made partitions must be constructed undersized to allow for positioning tolerances*

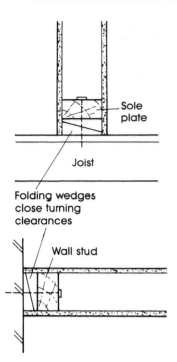

Figure 3.28
Use of folding wedges to take up positioning tolerances

Drilling and notching partitions

Service cables and pipes for water, gas and electricity are often concealed within stud partitions. In common with floor joists the positioning of holes and notches in studs has an effect on strength. It is recommended that holes and notches in studs should be kept to a minimum and conform to the following (Figure 3.29).

Holes of up to 0.25 of the stud's width, drilled on the centre line (neutral stress line) and located between 0.25 and 0.4 of the stud's height from either end are permissible. Adjacent holes should be separated by at least three times their diameter measured centre to centre.

Notches on either edge of the stud up to 0.15 of the stud's width and located up to 0.2 of the stud's height from either end are permissible.

> *example*
>
> The position and sizes for holes and notches in 100 mm width studs, 2400 mm in length are:
>
> **Holes** between 600 mm and 960 mm in from either end of the stud and up to 25 mm in diameter.
>
> **Notches** up to 480 mm in from each end and up to 15 mm deep.

Excessive drilling and notching outside these permissible limits will weaken the stud and may lead to failure. In addition, holes and notches should be kept clear of areas where the services routed through them are likely to get punctured by nails and screws, e.g. behind skirtings, dado rails and kitchen units etc., or metal plates can be fitted for protection.

Fire resistance, thermal insulation and sound insulation

Depending on the location of the partition it may be required to form a fire resisting and/or thermal insulating and/or sound insulating function. Specific requirements and methods of achieving them are controlled by the Building Regulations, which should be referred to for information as may be required. However, the general principles for achieving these functions are as follows (see Figure 3.30):

◆ Fire resistance can be increased by a double covering of plasterboard, the second outer layer being fixed so that the joints overlap those in the lower layer.
◆ Thermal insulation can be increased by filling the space between studs with either mineral wool or glass fibre quilt.
◆ Sound insulation can be increased by a discontinuous construction to avoid impact and vibration and the use of lightweight infilling material such as mineral or glass fibre quilt to absorb sound energy.

Figure 3.29
Positioning and protection of holes and notches to accommodate services

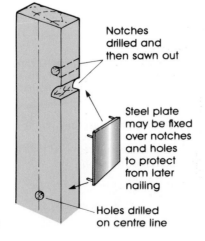

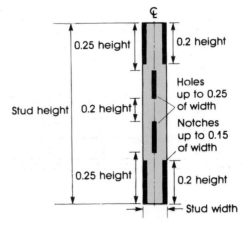

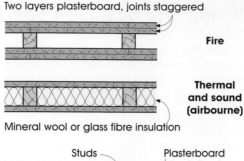

Two layers plasterboard, joints staggered

Fire

Thermal and sound (airbourne)

Mineral wool or glass fibre insulation

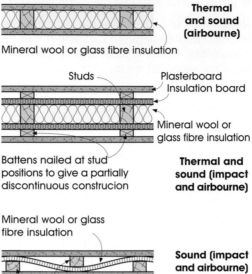

Studs   Plasterboard
                         Insulation board

Mineral wool or glass fibre insulation

Thermal and sound (impact and airbourne)

Battens nailed at stud positions to give a partially discontinuous construcion

Mineral wool or glass fibre insulation

Sound (impact and airbourne)

Staggered studs for discontinuous construction

Figure 3.30 *Fire, thermal and sound insulation of stud partition*

activity

Determine the number of 2400 mm long studs required to be spaced at 600 mm centres, for a partition between two walls 4500 mm apart.

Determine the total amount of timber required in metres run for the partition. Include head plate, sole plate, studs and three lines of noggins.

measuring up

11. State the reason why pre-made partitions are constructed shorter than the height of the room where they are to be fixed.

12. Produce a sketch to show a suitable means of jointing a sole plate at a return.

13. A stud partition wall is to be dry lined. Which face should be fixed outwards?

14. State the purpose of drilling or notching studs.

15. State the maximum size hole that should be drilled in a 100 mm wide stud.

16. Name the regulations that govern fire resistance, thermal insulation and sound insulation of partitions.

17. Name TWO methods of securing glazing beads to the lining of a borrowed light.

18. Determine the total length of timber in metres run required for the following partition:
    Eleven 50 mm × 75 mm studs 2400 mm long
    Head and sole plate 4100 mm long
    Three lines of noggins

Stairs

A stairway can be defined as a series of steps (combination of tread and riser) giving floor-to-floor access. Each continuous set of steps is called a flight. Landings are introduced between floor levels either to break up a long flight, giving a rest point, or to change the direction of the stair.

Straight-flight stairs

These run in one direction for the entire length. Figure 3.31 shows there are three different variations.

The flight that is closed between two walls (also known as a cottage stair) is the simplest and most economical to make. Its handrail is usually fixed either directly on to the wall or on brackets.

The flight fixed against one wall is said to be open on one side. This open or outer string is normally terminated and supported at either end by a newel post. A balustrade must be fixed to this side to provide protection. The infilling of this can be either open or closed and is usually capped by a handrail. Where the width of the flight exceeds 1 m, a wall handrail will also be required.

Where the flight is freestanding, neither side being against a wall, it is said to be open both sides. The open sides are treated in the same way as the previous flight.

Stair construction

The four main methods of constructing straight-flight stairs are illustrated in Figure 3.32:

◆ **Close string** – having parallel strings with the treads and risers being housed into their faces and secured by gluing and wedging.
◆ **Cut string** – having one or more strings that have been cut to conform to the tread and riser profile. Treads sit on the strings' horizontal cut portion and the risers are mitred to the vertical portion.
◆ **Open riser** – stairs with no risers, also termed open plan stairs. Treads are either housed or mortised into the strings, often with a ranch-style planked balustrade.
◆ **Alternating tread** – a narrow steeply pitched form of open riser stair used for access to a loft conversion in a domestic property, where there is insufficient space to accommodate a full size stair.

Stair terminology

Apron lining – the boards used to finish the edge of a trimmed opening in the floor.

Balustrade – the handrail and the infilling between it and the string, landing or floor. This can be called either an open or closed balustrade, depending on the infilling.

Baluster – the short vertical infilling members of an open balustrade.

Bull nose step – the quarter-rounded end step at the bottom of a flight of stairs.

Carriage – a raking timber fixed under wide stairs to support the centre of the treads and risers. Brackets are fixed to the side of the carriage to provide further support across the width of the treads.

Commode step – a step with a curved tread and riser normally occurring at the bottom of a flight.

Curtail step – the half-rounded or scroll-end step at the bottom of a flight.

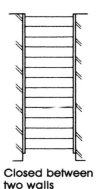

Closed between
two walls

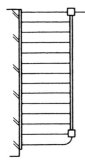

Against wall
open one side

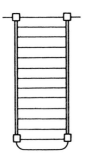

Free-standing
open both sides

Figure 3.31 *Straight-flight stairs*

First Fixing

Chapter 3

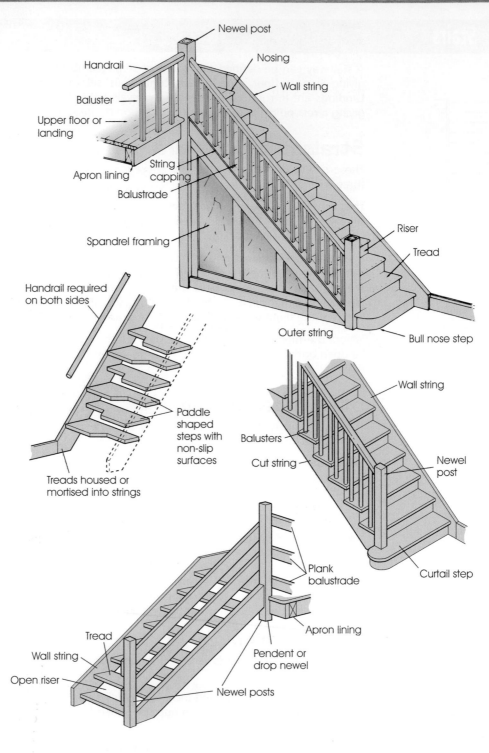

Figure 3.32

Straight flight stair
construction and terminology

Newel – the large sectioned vertical member at each end of the string. Where an upper newel does not continue down to the floor level below it is known as a pendant or drop newel.

Nosing – the front edge of a tread or the finish to the floorboards around a stairwell opening.

Riser – the vertical member of a step.

Spandrel – the triangular area formed under the stairs. This can be left open or closed in with spandrel framing to form a cupboard.

String – the board into which the treads and risers are housed or cut. They are also named according to their type, for example, wall string, outer string, close string, cut string and wreathed string.

Tread – the horizontal member of a step. It can be called a parallel tread or a tapered tread etc., depending on its shape.

Stair installation

did you know?

Coach screws are heavy-duty screws, which are inserted using a spanner rather than a screwdriver.

Stairs are normally delivered to site assembled as far as possible, but for ease of handling each flight will be separate. Its newels, handrail and balustrade are supplied loose, ready for on-site completion.

For maximum strength and rigidity the stairs should be fixed as shown in Figures 3.33 to 3.37. The top newel is notched over the landing or floor trimmer and either bolted or coach screwed to it. The lower newel should be carried through the landing or floor and bolted to the joists. The lower newel on a solid ground floor can be fixed by inserting a steel dowel partly into the newel and grouting this into the concrete.

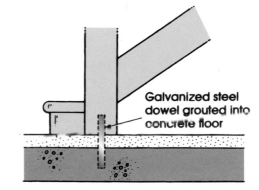

Figure 3.34 *Alternative fixing of lower newel*

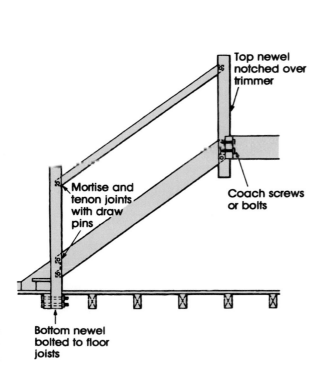

Figure 3.33 *Fixing outer string-newels and handrail*

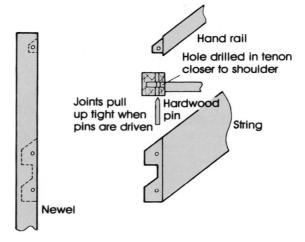

Figure 3.35 *String handrail to newel joints*

First Fixing Chapter 3

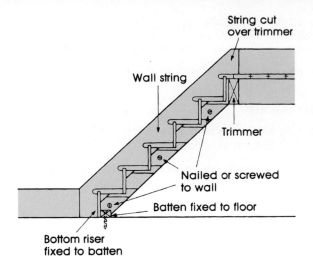

Figure 3.36 *Fixing wall string*

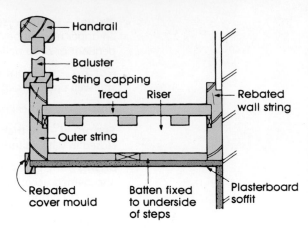

Figure 3.37 *Section across flight open one side*

The outer string and handrail are mortised into the newels at either end. With the flight in position, these joints are glued and then closed up and fixed using hardwood draw pins. The wall string is cut over the trimmer at the top and cut-nailed or screwed to the wall from the underside. The bottom riser of a flight may be secured by screwing it to a batten fixed to the floor.

Figure 3.38 shows how the trimmer around the stairwell opening is finished with an apron lining and nosing.

Where the width of the stair exceeds about 1 m, a carriage may be fixed under the flight to support the centre of the treads and risers. To securely fix the carriage it is birdsmouthed at both ends, at the top over the trimmer and at the bottom over a plate fixed to the floor. Brackets are nailed to alternate sides of the carriage to provide further support across the width of the treads. See Figure 3.39.

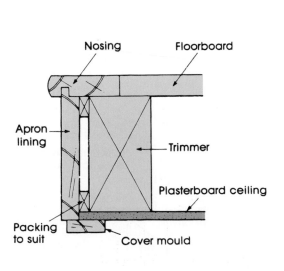

Figure 3.38 *Landing detail*

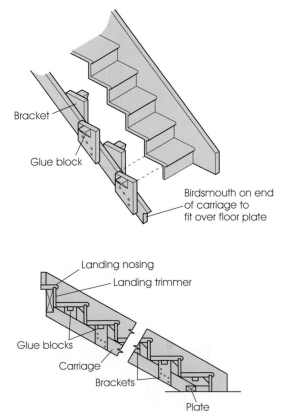

Figure 3.39 *Carriage on wider stairs*

First Fixing

Chapter 3

Balustrades and wall handrails

The main types of balustrade infill used in domestic property are either timber balusters or ranch-style planking.

Timber balusters

Timber balusters may be either stub tenoned into the string or fitted into a groove run in the string capping. A distance piece may be fitted to maintain the spacing where they are fitted into a groove. At the upper end balusters may be located in a groove or be stub tenoned into the handrail, as illustrated in Figure 3.40. The planks of ranch-style balustrades may be tenoned into the newels or screwed directly to the face of the newels.

A half newel can be used to support the hand and balustrade where it returns around a landing, see Figure 3.41.

Wall handrails

Handrails on at least one side fixed between 900 mm and 1000 mm above the pitch line are required on all flights. In open stairs this may be the handrail of the balustrade. Wall handrails are required on one side of flights that are closed between two walls and other flights, which are wider than 1 m. They may be fixed directly to the wall with screws and plugs or stand clear of the wall on brackets as illustrated in Figure 3.42.

Protection of completed work

After a new staircase has been installed, a short period spent taking measures to prevent damage during subsequent building work saves much more than it costs.

False treads made from strips of hardboard or plywood, as shown in Figure 3.43, are pinned on to the top of each step. The batten fixed to the strip ensures the nosing is well protected. On flights that are to be clear finished the false treads should be held in position with a strong adhesive tape, as pin holes would not be acceptable.

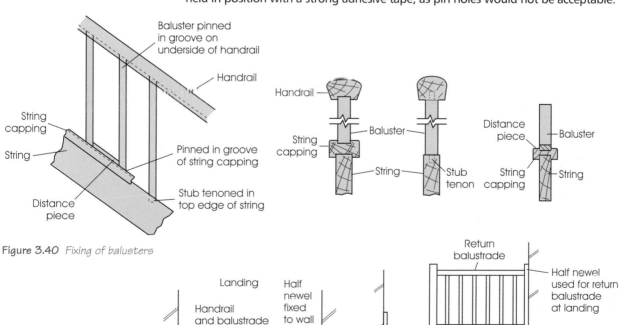

Figure 3.40 Fixing of balusters

Figure 3.41 Return balustrades at landings

First Fixing Chapter 3

Strips of hardboard or plywood are also used to protect newel posts. These can be either pinned or taped in position depending on the finish (see Figure 3.44). Adequate protection of handrails and balustrades can be achieved by wrapping them in corrugated cardboard held in position with adhesive tape.

Adequate protection of handrails and balustrades can be achieved by wrapping them in corrugated cardboard or bubble wrap held in place with adhesive tape as shown in Figure 3.45.

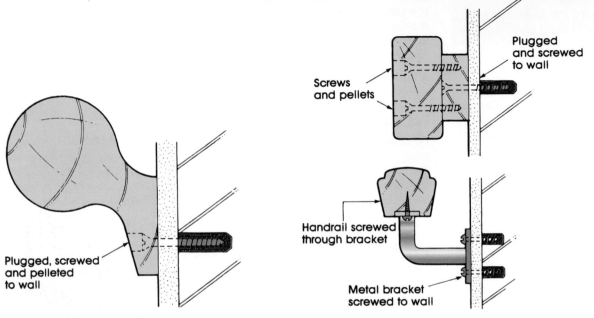

Screws and pellets

Plugged and screwed to wall

Handrail screwed through bracket

Metal bracket screwed to wall

Plugged, screwed and pelleted to wall

Figure 3.42 *Handrail sections*

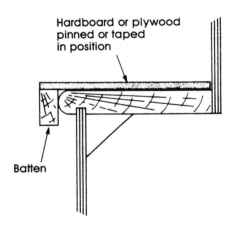

Hardboard or plywood pinned or taped in position

Batten

Figure 3.43 *Temporary protection of treads*

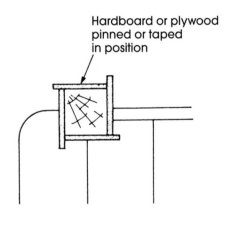

Hardboard or plywood pinned or taped in position

Figure 3.44 *Temporary protection of newels*

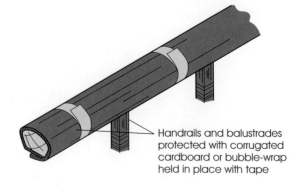

Handrails and balustrades protected with corrugated cardboard or bubble-wrap held in place with tape

Figure 3.45 *Temporary protection of handrails and balusters*

Using the extract from a stair manufacturer's brochure below, determine the number of balusters (spindles) required to construct the balustrade shown in the example.

S T A I R P A R T S

HELPFUL HINTS

When calculating the number of 32mm spindles needed on the staircase itself, allow 2 spindles per tread and 1 per tread where there is a newel.

On the landing, to calculate the number of spindles required, (X) simply use the following formula:

$$X = \frac{\text{Horizontal distance (in mm) on landing between newels}}{112}$$

Example

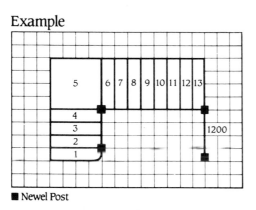

■ Newel Post

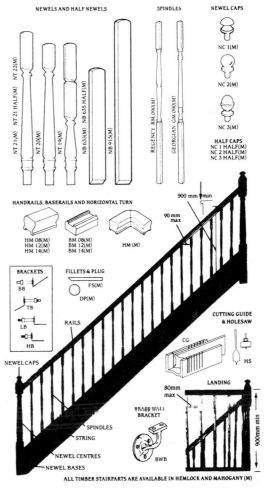

Reproduced with the permission of Richard Burbidge Ltd, Whittington Road, Oswestry, Shropshire SY11 1HZ. For further information please telephone 01691 655131.

19. List FIVE components that may be used in a flight of stairs.

20. Produce a sketch to show how a newel post may be fixed to the landing trimmer.

21. Name the vertical member used to provide support to a handrail and infill an open balustrade.

22. List THREE measures that can be taken after stair installation to prevent damage to a flight during building work.

First Fixing Chapter 3

Frames and linings

Frames and linings terminology

Frame – an assembly of components to form an item of joinery, such as a door or window; a structural framework of columns and beams or panels in steel reinforced concrete or timber.

Lining – the thin covering to door or window reveals; sheet material used to cover wall surfaces.

Door frame – the surround on which an external door or internal door is hung consisting of two jambs, a head and sometimes a threshold and transom; normally with stuck-on solid stops and of a bigger section than door linings. (See Figure 3.46.)

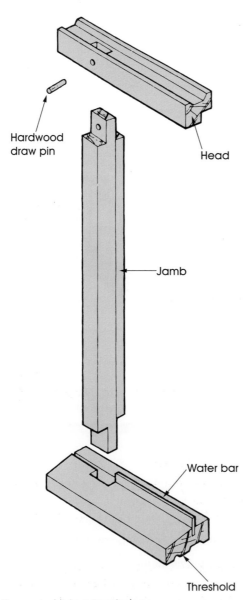

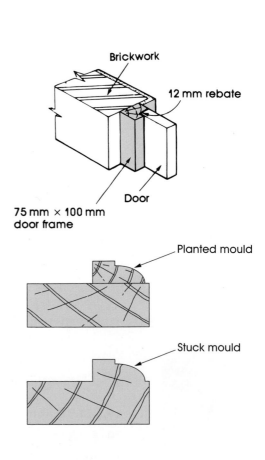

Figure 3.46 *Frames and linings terminology*

Door lining – the surround on which mainly internal doors are hung, normally of a thinner section than door frames and often have planted stops. The main difference between door frames and door linings is that linings cover the full width of the reveal in which they are fixed from wall surface to wall surface, whereas frames do not (Figure 3.47).

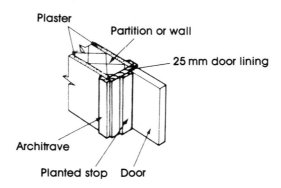

Figure 3.47 *Door lining*

Window frame – the part of a window that is fixed into the wall opening and receives the casements or sashes (Figure 3.48).

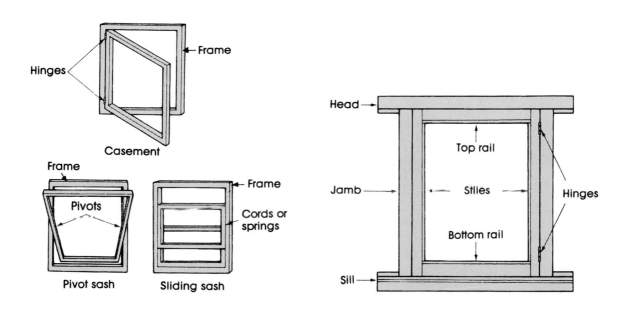

Figure 3.48 *Window terminology*

Frames

Frames can be either 'built in' or 'fixed in'.

◆ 'Built-in' frames are fixed into a wall or other element by bedding in mortar and surrounded with the walling components.
◆ 'Fixed-in' frames are inserted into a ready formed opening after the main building process.

'Built-in' frames

The majority of frames are 'built-in' by the bricklayer as the brickwork proceeds. Prior to this the frame has to be accurately positioned, plumbed, levelled and temporary strutted by the carpenter. Door frames are normally built into the brickwork as the work proceeds. Temporary struts are used to hold the frame upright. The foot of the door frame jambs, in the absence of a threshold, are

did you know?

You should close one eye when sighting-in to eliminate bifocal vision. The image in one eye is somewhat different from that in the other, so a more defined view is obtained with one eye closed. Try closing one eye at a time. The best view is achieved with your dominant eye.

held in position by galvanized metal dowels, which are drilled into the end of the jambs and are grouted into the concrete. This is shown in Figure 3.49. Temporary braces and distance pieces are fixed to the frame, in order to keep it square and the jambs parallel during the 'building-in' process. The vertical positioning of external door and window frames can be achieved with the use of a storey rod.

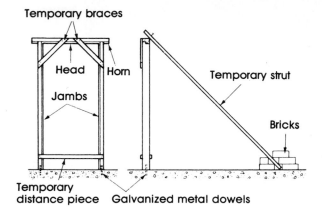

Figure 3.49 'Building-in' a frame

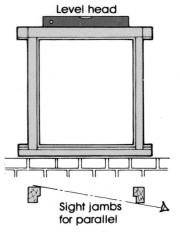

A frame's head should be checked for level, and packed up as required; frames with thresholds are normally bedded level using bricklayer's mortar (see Figure 3.50).

Jambs should be plumbed from the face. It is standard practice to plumb and fix the first using a spirit level. The other is then sighted parallel: stand to the side of the frame, close one eye, sight the edge of the plumbed jamb with the edge of the other and adjust if required until both jambs are parallel.

As the brickwork proceeds galvanized metal frame cramps, as shown in Figure 3.51, should be screwed to the back of the jambs and built into the brickwork. Three or four cramps should be evenly spaced up each jamb.

Figure 3.51 Attaching frame cramps

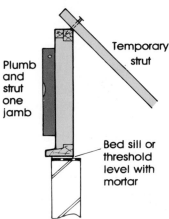

Figure 3.50 Positioning 'built-in' frames

The horns of the frame should be cut back as shown before 'building in', rather than cut off flush (see Figure 3.52). The horns will then help to fix the frame in position. After trimming the horns it is essential that the cut ends are treated with preservative in order to reduce the possibility of timber rot. This can be carried out by applying two brush flood coats of preservative.

Storey height frames may be used for internal door openings in thin blockwork partitions (Figure 3.53). The jambs and head that make up the frame are grooved out on their back face to receive the building blocks. The storey frame should be

Chapter 3 First Fixing

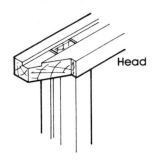

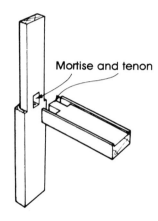

Figure 3.52 *Horn cut back ready for 'building-in'*

Figure 3.53 *Joint detail (storey height frame)*

fixed in position, at the bottom to the wall plate and at the top of the joists, before the blocks are built up (see Figure 3.54). The jambs above the head are cut back to finish flush with the blockwork. As with other frames one jamb should be fixed plumb and the other sighted to it.

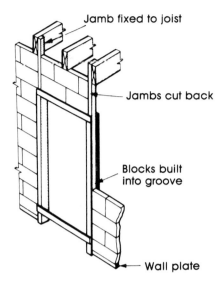

Figure 3.54 *'Building-in' a storey height frame*

Profiles

Where door and window frames are not available at the time when the wall is being built, or they are made from easily damaged materials such as polished hardwood or plastic, temporary profile frames or templates are used, leaving the frames to be fixed at a later stage.

The temporary profiles can be made using 50 × 75 mm softwood, secured with plywood corner gussets as illustrated in Figure 3.55. In order to allow for a fixing tolerance the profile should be made about 6 mm wider than the width of the frame. The height of the profile is made the same as the frame; a fixing tolerance is achieved by placing the frame on packers, which also aids the profiles removal once the brickwork has set.

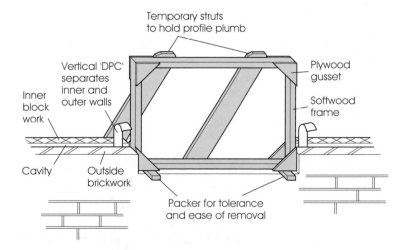

Figure 3.55 *Profile frame for windows that are to be fixed at a later stage*

First Fixing

Chapter 3

'Fixed-in' frames

'Fixed-in' frames are sometimes fixed to prepared openings. This applies mainly to expensive hardwood frames and is to protect them from possible damage or discoloration during the building process. In addition, frames that were not available during the building process or replacement frames will have to be 'fixed-in'.

Horns on 'fixed-in' frames are not required as a fixing and should be sawn off flush with the back of the jambs. Remember to treat the cut ends with preservative.

Place the frame in the prepared opening, temporarily holding it in position with the aid of folding wedges (Figure 3.56). Check the head or sill for level and adjust the wedges as required.

Plumb one jamb and 'sight in' the other; adjust the wedges if required.

did you know?

Cut-back horns must be treated with preservative before 'building-in'.

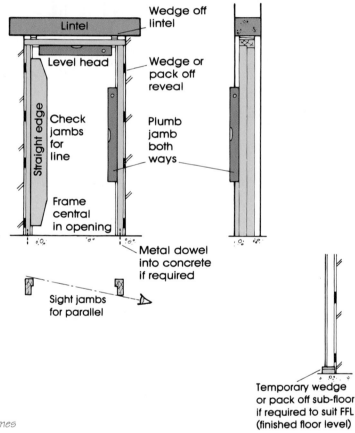

safety tip

Remember, when using preservatives, paints or other chemicals, always follow the manufacturer's safety instructions concerning their use and the appropriate PPE to be worn.

Figure 3.56 *Positioning 'fixed-in' frames*

Figure 3.57 shows how to fix the frame to the wall in the following ways:

◆ Nailing using cut nails into the blockwork or brickwork mortar joints or masonry nails into the actual brick.
◆ Screwing plastic plugs and screws through the jamb. Screwheads in softwood frames may be countersunk below the surface and filled. Screwheads in hardwood frames should be concealed by counter boring and pelleting.
◆ Metal plates used as fixing lugs screwed at intervals to the back of jamb before the frame is put in the opening. The lugs are screwed and plugged to the brick or block reveals.
◆ Frame anchors – proprietary fixing consisting of a metal or plastic sleeve and matching screw. The jamb and reveal are drilled out to suit the sleeve, which is inserted in position and screwed up tight.

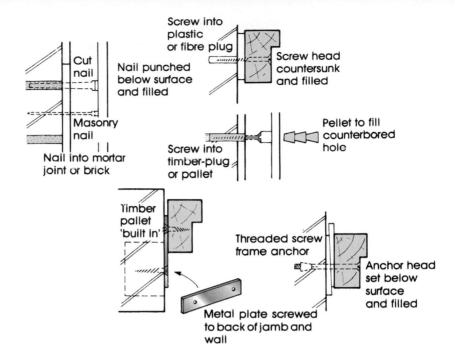

Figure 3.57 *Methods of fixing*

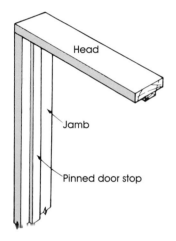

Figure 3.58 *Plain lining*

Linings

Plain linings – (Figures 3.58 and 3.59) consist of two plain jambs and a plain head joined together using a bare-faced tongue and housing. The pinned stop is fixed around the lining after the door has been hung.

Rebated linings – (Figures 3.60 and 3.61) are used for better quality work. They consist of two rebated jambs and a rebated head. The rebate must be the correct width so that when the door is hung it finishes flush with the edges of the lining.

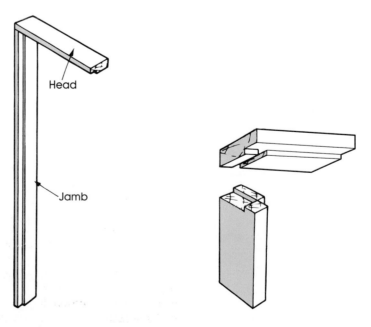

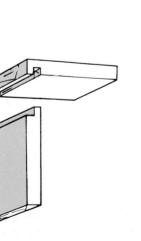

Figure 3.59 *Joint detail (plain lining)*

Figure 3.60 *Rebated lining*

Figure 3.61 *Joint detail (rebated lining)*

First Fixing

Chapter 3

Fixing linings

The opening in the wall to receive the lining is normally formed while the wall is being built and the lining is fixed at a later stage.

Fixings may be:

◆ Nailing to twisted wooden plugs (see sequence of operations).
◆ Nailing and screwing to timber pallets that have been built into the brick joints by the bricklayer or into the door stud of a stud partition. Folding wedges are used as packings down the sides of the jambs.
◆ Nailing directly into the blockwork reveal or brickwork mortar joint. Folding wedges will be required as packing.
◆ Using plugs and screws or other proprietary fixing.

The sequence of operations to fix a lining (using twisted timber plugs) is shown in Figure 3.62:

1. Assemble lining. This is normally done by skew nailing through the head into the jambs.
2. Fix a distance piece near the bottom of the jambs and, when required, diagonal braces at the head.
3. Rake out brickwork joints and plug (see Figure 3.63). There should be at least four fixing points per jamb. Omit this stage if the bricklayer has 'built-in' wooden pallets or pads into the brickwork.
4. Offer lining into opening and mark where the plugs need to be trimmed. The plugs should project equally from both reveals.
5. Cut the plugs and check the distance with a width rod. The ends of the plugs should be in vertical alignment. Check with a straight edge and spirit level.
6. Fix lining plumb and central in the opening by nailing or screwing through the jambs into the plugs. Before finally fixing check head for level, wedge off lintel, ensure the lining is out of wind: check by sighting through the jambs. When fixing to unplastered walls, check adjacent linings and wall surfaces are lined up.
7. Ensure lining jambs are packed up off a concrete sub floor if required to suit the finished floor level (FFL).

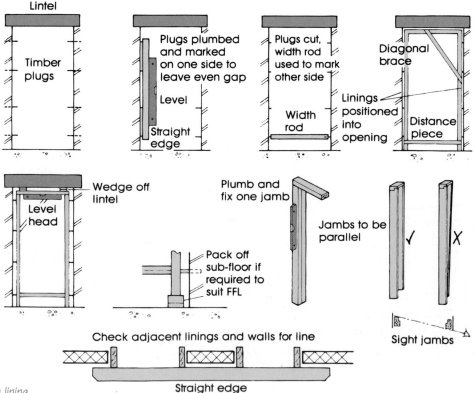

Figure 3.62 *Fixing a lining*

Weather sealing of external frames

All external frames should have the gap between them and the wall sealed with a silicone or acrylic sealant prior to occupation of the building. This is to prevent the penetration of wind, rain and insects into the gap and to prevent them from possibly finding their way into the building.

Frame sealants are normally supplied in cartridges and applied using a skeleton gun. The sealant should be forced as far back into the gap as possible, finally finishing off with a narrow angled face bead (see Figure 3.66). Any slight unevenness in the face bead can be smoothed off using a small paintbrush dipped in water.

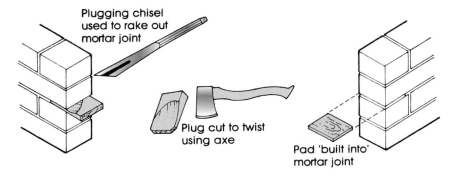

Figure 3.63 *Plugs or pads used for fixing lining*

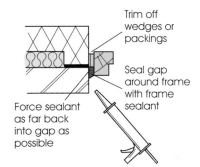

Figure 3.64 *Weather sealing of external frames*

Window boards

Window boards are normally used to finish the top of the wall internally at sill level. They may be formed from solid timber, plywood, MDF (medium density fibreboard) or plastic and should be fitted at the first fixing stage before the internal walls have been plastered. Where the window board fits up to a timber window, it is normally tongued into the sill groove. If the window is metal or plastic (PVCu) the window board is only butted to the sill with the gap being filled with caulking or covered with a small bead (see Figure 3.65).

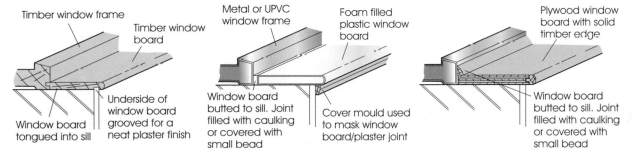

Figure 3.65 *Window board details*

First Fixing

Chapter 3

Many window boards have a groove on the underside; this provides a finish for the plaster and masks any gap, which may result later from shrinkage. Alternatively, a cover mould may be used to mask the joint.

Cutting and fixing window boards

See Figure 3.66:

1. Cut a length of board about 100 mm longer than the sill.
2. Mark out each end and cut out a portion of the board to fit the window reveal.
3. Return the front edge nosing profile around the ends using a plane and finish with an abrasive paper.
4. Use an offcut of window board and a boat level to check for front to back level. Packing may be required under the board where it is not level; proprietary plastic shims available in a range of thicknesses are the best. However a piece of hardboard or several pieces of damp-proof material will also do the job. These should be positioned about 50 mm from each end, with intermediates spaced at about 450 mm centres.
5. Place the window board in position and fix the front edge to the wall, using cut nails, masonry nails or plugs, screws and pellets. Fixings should preferably go through any packers so the board is not pulled down out of line and also to ensure they are not misplaced later.

Chapter 3   First Fixing

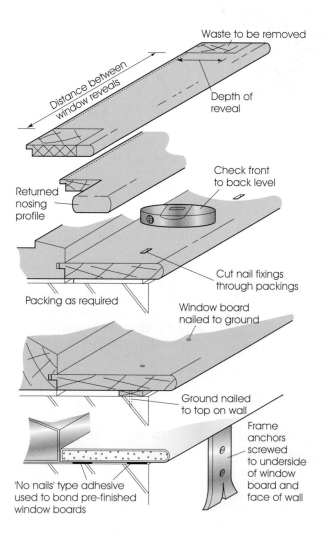

Figure 3.66 *Fixing window boards*

Alternative fixing methods

A batten or ground may be pre-fixed to the wall at the correct level for the window board, using cut or masonry nails, the window board in turn being fixed directly to the ground using oval nails. Plastic and other pre-finished window boards may be bedded and fixed to the wall using beads of a gap-filling 'no nails' type adhesive, or frame anchors screw-fixed to the underside of the window board and face of the wall.

activity

Refer to the floor plans, range drawing and schedules shown in Figures 4.23, 4.24 and 4.25.

Determine in metres run the amount of 38 mm × 125 mm and 38 mm × 100 mm required to construct the door linings. Use 10% for a cutting allowance.

measuring up

23. Produce a sketch to show the difference between a door frame and a door lining.

24. Explain the difference between 'built-in' and 'fixed-in' frames and state an occasion where EACH might be used.

25. Explain the reason why a steel dowel may be included in the base or foot of newel post and door frame jambs.

26. Describe the procedure for cutting and fixing a timber window board.

27. Explain how to 'sight-in' the jambs of a frame.

28. State the purpose of profile frames.

29. List THREE forms of fixing that can be used to secure frames or linings.

30. State the purpose of using frame sealant.

First Fixing

Chapter 3

Second Fixing

This chapter is intended to provide the reader with an overview of second fixing work. Its contents are assessed in the NVQ Unit No. VR 10 Install Second Fixing Components.

In this chapter you will cover the following range of topics:

◆ doors;
◆ doors and ironmongery;
◆ finishing trim;
◆ encasing services;
◆ kitchen units and fitments;
◆ panelling and cladding.

What's required in VR 10 Install Second Fixing Components?

To successfully complete this unit you will be required to demonstrate your skill and knowledge of the following second fixing elements:

◆ door hanging;
◆ mouldings;
◆ units and fitments;
◆ cladding.

You will be required practically to:

◆ install side hung doors;
◆ install skirting and architraves;
◆ install a range of door ironmongery;
◆ install wall and floor units or fitments;
◆ install cladding.

Note: Except for cladding, for the successful completion of all other practical tasks evidence must be work based.

 Doors

Doors may be classified by their method of construction: panelled, glazed, matchboarded, flush, fire resistant etc. (Figure 4.1), and by their method of operation: swinging, sliding and folding.

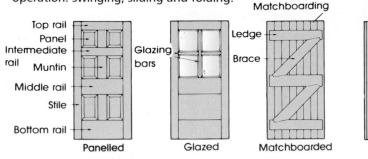

Figure 4.1 *Doors*

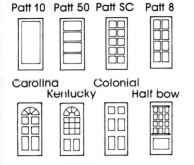

Internal doors

Framed
Patt SA   Colonial   Patt
           pine   2XGG

Flush
White-faced Plywood   Sapele
hardboard   (painted) (polished)

Sizes available
610 mm × 1981 mm × 35 mm
686 mm × 1981 mm × 35 mm
762 mm × 1981 mm × 35 mm
865 mm × 1981 mm × 35 mm

External doors

Patt 10 Patt 50 Patt SC Patt 8

Carolina   Colonial
  Kentucky   Half bow

Sizes available
762 mm × 1981 mm × 44 mm
835 mm × 1981 mm × 44 mm
813 mm × 2032 mm × 44 mm

Figure 4.2 *Extract from a door manufacturer's door list showing the range of stock size doors*

Methods of construction

Panelled doors

Panelled doors have a frame made from solid timber rails and stiles, which are jointed using either dowels or mortise and tenon joints. The frame is either grooved or rebated to receive two or more thin plywood or timber panels. Interior doors are thinner than exterior doors.

Glazed doors

Glazed doors are used where more light is required. They are made similar to panelled doors except that glass replaces one or more of the plywood or timber panels. Glazing bead is used to secure the glass into its glazing rebates. Glazing bars may be used to divide large glazed areas.

Matchboarded doors

Matchboarded doors are used mainly externally for gates, sheds and industrial buildings. They are simply constructed from matchboarding, ledges and braces clench nailed together. The bottom end of the braces must always point towards the hanging edge of the door to provide the required support. Framed matchboarded doors constructed with the addition of stiles and rails are used where extra strength is required.

Flush doors

Flush doors are made with outer faces of plywood or hardboard. Internal doors are normally lightweight, having a hollow core, solid timber edges and blocks that are used to reinforce hinge and lock positions. New flush doors will have one edge marked 'LOCK' and the other 'HINGE'; these must be followed. External and fire-resistant flush doors are much heavier, as normally they have a solid core of either timber strips or chipboard.

Fire resisting doors

Fire resisting doors are mainly constructed as solid-core flush doors. The main function of this type of door is to act as a barrier to a possible fire by providing the same degree of protection as the element in which it is located. They should prevent the passage of smoke, hot gases and flames for a specified period of time. This period of time will vary depending on the relevant statutory regulations and the location of the door. Fire doors are not normally purpose made, as they must have approved fire resistance certification. It is advantageous to use proven proprietary products. Oversize fire doors 'blanks' are available for cutting down to size if required to suit specific situations.

Grading of doors

The quality of a door will obviously affect its useful life (length of time that it is able to give satisfactory service). The majority of doors are therefore graded as being of either an internal or external quality.

Door sizes

All mass-produced doors may be purchased from a supplier in a range of standard sizes as shown in Figure 4.2. Special sizes or purpose-made designs are normally available to order from suppliers with joinery shop contacts.

In general internal doors have a finished thickness of either 35mm or 40mm. The thickness of external doors and fire doors is normally increased to 44mm, in order to withstand the extra stresses and strains to which they are subjected.

While the type of adhesive used in manufacture is a significant factor to be considered when making purpose-made doors, in practice it is of little significance when using standard doors as most manufacturers use a synthetic adhesive for both grades. In addition, external doors should be preservative treated against fungal decay.

Methods of operation

Swinging doors – Side hung on hinges is the most common means of door operation. It is also the most suitable for pedestrian use and the most effective for weather protection, fire resistance, sound and thermal insulation. See Figure 4.3.

Sliding doors – are mainly used either to economise on space where it is not possible to swing a door, or for large openings which would be difficult to close off with swinging doors.

Folding doors – are a combination of swinging and sliding doors. They can be used as either movable internal partitions to divide up large rooms, or alternatively as doors for large warehouses and showroom entrances.

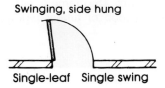

Swinging, side hung

Single-leaf   Single swing

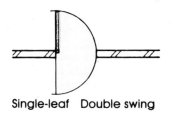

Single-leaf   Double swing

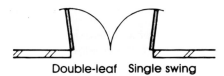

Double-leaf   Single swing

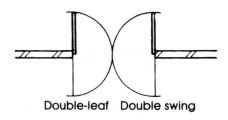

Double-leaf   Double swing

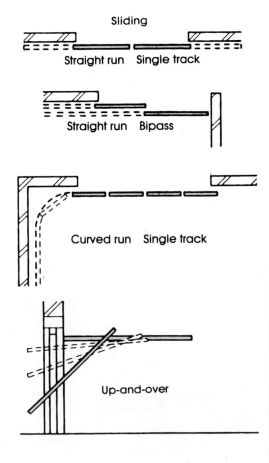

Sliding

Straight run   Single track

Straight run   Bipass

Curved run   Single track

Up-and-over

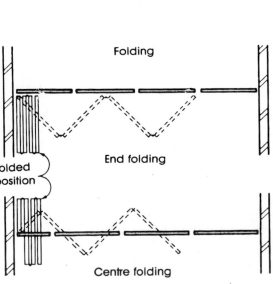
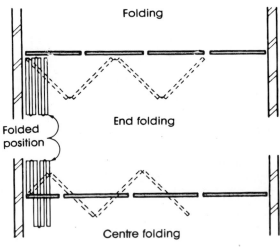

Folding

End folding

Folded position

Centre folding

Figure 4.3 *Methods of door operation*

Door ironmongery

Door ironmongery is also termed 'door furniture' and includes hinges, locks, latches, bolts, other security devices, handles and letter or postal plates. The hand of a door is required in order to select the correct items of ironmongery. Some locks and latches have reversible bolts, enabling either hand to be adapted to suit the situation.

Traditionally, this has always been done by viewing the door from the hinge knuckle side; if the knuckles are on the left the door is left handed, whereas if the knuckles are on the right, the door is right handed. Doors may also be defined as either clockwise or anti-clockwise closing when viewed from the knuckle side. When you are ordering ironmongery, simply stating left-hand or right-hand, clockwise or anti-clockwise can be confusing, as there may be variations between manufacturers and suppliers. The standard way now to identify handing is to use the following coding (and see Figure 4.4):

◆ 5.0 for clockwise closing doors and indicating ironmongery fixed to the opening face (knuckle side)
◆ 5.1 for clockwise closing doors and indicating ironmongery fixed to the closing face (non-knuckle side)
◆ 6.0 for anti-clockwise closing doors and indicating ironmongery fixed to the opening face
◆ 6.1 for anti-clockwise closing doors and indicating ironmongery fixed to the closing face.

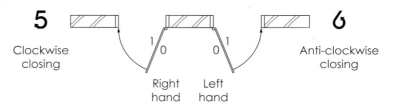

Figure 4.4 *Method for stating door handing*

Hinges

Hinges are available in a variety of materials: pressed steel is commonly used for internal doors and brass for hardwood and external doors. Do not use steel hinges on hardwood or external doors because of rusting and subsequent staining problems. Do not use nylon, plastic or aluminium hinges on fire-resistant doors because they melt at fairly low temperatures.

Butt hinge

Butt hinge is a general-purpose hinge suitable for most applications. As a general rule the leaf with the greatest number of knuckles is fixed to the door frame.

Flush hinges

Flush hinges can be used for the same range of purposes as a butt hinge on both cabinets and full-size room doors. They are only really suitable for lightweight doors, but they do have the advantage of easy fitting, as they do not require 'sinking in'.

Loose pin butt hinges

Loose pin butt hinges enable easy door removal by knocking out the pins. For security reasons they should not be used for outward opening external doors.

Lift-off butt hinges

Lift-off butt hinges also enable easy door removal, the door being lifted off when in the open position. These are available as right-handed or left-handed pairs. Each pair consists of one long pin hinge, which is fitted as the lower hinge. The upper hinge has a slightly shorter pin, to aid repositioning the door.

Washered butt hinges

Washered butt hinges are used for heavier doors, to reduce knuckle wear and prevent squeaking.

Parliament hinges

Parliament hinges have wide leaves to extend knuckles and enable doors to fold back against the wall clearing deep architraves etc.

Rising butts

Rising butts are designed to lift the door as it opens to clear obstructions such as mats and rugs. They also give a door some degree of self-closing action. In order to prevent the top edge of the door fouling in the frame as it opens and closes, the top edge must be eased, as shown in Figure 4.5. The hand of the door must be stated when ordering this item, as they cannot be reversed (i.e. they cannot be altered to suit either hand of door).

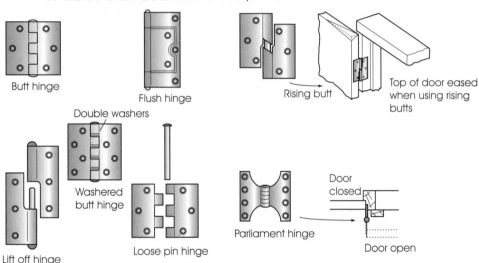

Figure 4.5 *Range of hinges for doors*

Butt hinge

Flush hinge

Rising butt

Top of door eased when using rising butts

Double washers

Washered butt hinge

Lift off hinge

Loose pin hinge

Parliament hinge

Door closed

Door open

did you know?

Narrow style vertical mortise lock/latches are also known as sash lock/latches.

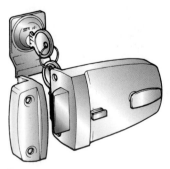

Figure 4.6 *Cylinder rim latch*

Locks and latches

Cylinder rim latches

Cylinder rim latches are mainly used for entrance doors to domestic property, but, as they are only a latch, provide little security on their own (Figure 4.6). When fitted, the door can be opened from the outside with the use of a key and from the inside by turning the handle. Some types have a double locking facility that improves their security.

Mortise deadlock

A mortise deadlock provides a straightforward key-operated locking action and is often used to provide additional security on entrance doors where cylinder rim latches are fitted (Figure 4.7). They are also used on doors where simple security is required, e.g. storerooms. The more levers a lock has the better it is; a five-lever lock provides greater security than one with three levers.

Mortise latch

A mortise latch is used mainly for internal doors that do not require locking (Figure 4.8). The latch that holds the door in the closed position can be operated from either side of the door by turning the handle.

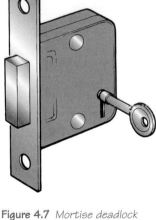

Figure 4.7 *Mortise deadlock*

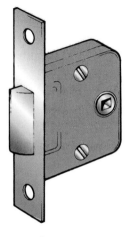

Figure 4.8 *Mortise latch*

Mortise lock/latch

A mortise lock/latch is available in the two types shown in Figure 4.9. The horizontal one is little used nowadays because of its length, which means that it can only be fitted to substantial doors. The vertical type is more modern and can be fitted to most types of doors. It is often known as a narrow-stile lock/latch. Both types can be used for a wide range of general purpose doors in various locations. They are, in essence, a combination of the mortise deadlock and the mortise latch. Further variations are the Euro pattern, which uses a cylinder lock to operate the dead bolt, and bathroom privacy patterns, which use a turn button on the inside of the door.

Rebated mortise lock/latch

A rebated mortise lock/latch should be used when fixing a lock/latch in double doors that have rebated stiles. The front end of this lock is cranked to fit the rebate on the stiles (see Figure 4.10).

Knobset

A knobset consists of a small mortise latch and a pair of knob handles that can be locked with a key, so that it can be used as a lock/latch in most situations, both internally and externally (Figure 4.11). Knobsets can also be obtained without the lock in the knob for use as a latch only.

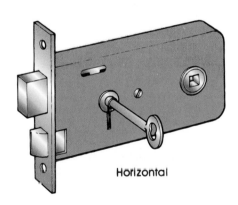

Horizontal

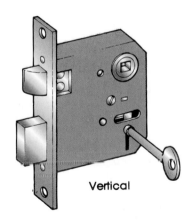

Vertical

Figure 4.9 *Mortise lock/latches*

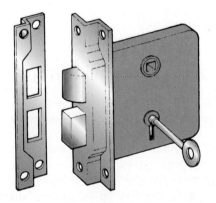

Figure 4.10 *Rebated mortise lock/latch*

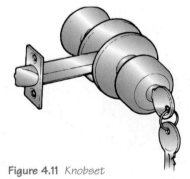

Figure 4.11 *Knobset*

Figure 4.12 *Knob furniture*

Figure 4.13 *Escutcheon plate*

Figure 4.14 *Lever furniture*

Figure 4.16 *Flush bolt*

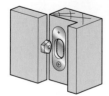

Figure 4.17 *Hinge bolt*

Knob furniture

Knob furniture is for use with the horizontal mortise lock/latch (Figure 4.12). It should not be used with the vertical type as hand injuries will result.

Keyhole escutcheon plates

Keyhole escutcheon plates are used to provide a neat finish to the keyhole of both deadlocks and horizontal mortise lock/latches (Figure 4.13).

Lever furniture

Lever furniture is available in a wide range of patterns, for use with mortise latches and mortise lock/latches (Figure 4.14).

Barrel and tower bolts

Barrel and tower bolts are used on external gates and doors to secure them from the inside. Two bolts are normally used, one near the top of the door and the other near the bottom (Figure 4.15).

Flush bolt

A flush bolt is flush fitting and therefore requires recessing into the timber (see Figure 4.16). It is used for better quality work on the inside of external doors to provide additional security and also on double doors and French windows to bolt one door in the closed position. Two bolts are normally used, one at the top of the door and the other at the bottom.

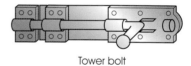

Barrel bolt

Tower bolt

Figure 4.15 *Barrel and tower bolts*

Hinge bolts

Hinge bolts help to prevent a door being forced off its hinges (see Figure 4.17). They provide increased security particularly on outward opening doors where the hinge knuckle pin is vulnerable.

Mortise rack bolts

Mortise rack bolts are fluted key operated dead bolts (see Figure 4.18). These are mortised into the edge of the door at about 150 mm from either end.

Security chains

Security chains can be fixed on front entrance doors, the slide to the door and the chain to the frame (Figure 4.19). When the chain is inserted into the slide, the door will only open a limited amount until the identity of the caller is checked.

Ironmongery positioning

Hinge size and position

◆ Lightweight internal doors are normally hung on one pair of 75 mm hinges.
◆ Glazed, half-hour fire resistant and other heavy doors need one pair of 100 mm hinges.
◆ All external doors and one-hour fire resistant doors need three (one-and-a-half pairs) of 100 mm hinges.

Figure 4.18 *Mortise rack bolt*

Figure 4.19 *Security chain*

The standard door hinge positions are illustrated in Figure 4.20.

For flush doors these are 150 mm down from the top, 225 mm up from the bottom and the third hinge where required, positioned centrally to prevent warping, or towards the top for maximum weight capacity. On panelled and glazed doors the hinges are often fixed in line with the rails to produce a more balanced look. However, slight adjustment may be required to avoid the end grain of wedges.

Other door furniture positions

The position of door ironmongery or furniture will depend on the type of door construction, the specification and the door manufacturer's instructions. Figure 4.21 illustrates the recommended fixing height for various items.

◆ The standard position for mortise locks and latches is 990 mm from the bottom of the door to the centre line of the lever or knob furniture spindle. However, on a panelled door with a middle rail, locks/latches may be positioned centrally in the rail's width.
◆ Cylinder rim latches are positioned in the door's stile between 1200 mm and 1500 mm from the bottom of the door and the centre line of the cylinder.
◆ Before fitting any locks/latches the width of the door stile should be measured to ensure the lock/latch length is shorter than the stile's width, otherwise a narrow stile lock may be required.
◆ On external doors using both a cylinder rim latch and a mortise dead lock, the best positions for security are one-third up for the dead lock and one-third down from the top for the cylinder rim latch.
◆ Security chains are best positioned near the centre of the door in height.
◆ Hinge bolts should be positioned just below the top hinge and just above the bottom hinge.
◆ Letter plates are normally positioned centrally in a door's width and between 760 mm and 1450 mm from the bottom of the door to the centre line of the plate. Again on panelled doors, letter plates may be positioned centrally in a rail and sometimes even vertically in a stile.
◆ Information signs are normally positioned centrally in a door's width and 1500 mm from the bottom of the door to the centre of the sign.
◆ Kicking plates are fixed to the bottom face of the door for protection, keeping an even margin. Push or finger plates are positioned near the closing edge of a door at a height of 1200 mm to the centre of the plate. On panelled and glazed doors, they should be fixed centrally in the stile's width.

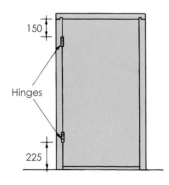

did you know?

You should always read and follow the manufacturer's instructions when fitting ironmongery to ensure its correct positioning and operation.

Second Fixing

Chapter 4

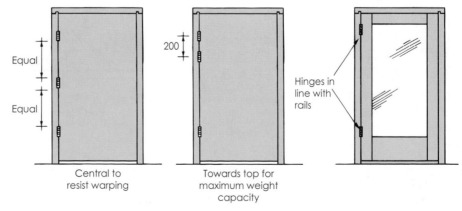

Figure 4.20 *Hinge positioning*

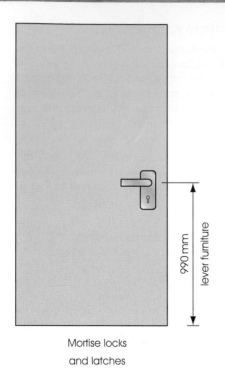

990 mm

lever furniture

Mortise locks
and latches

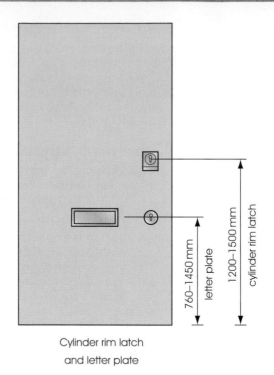

760–1450 mm

letter plate

1200–1500 mm

cylinder rim latch

Cylinder rim latch
and letter plate

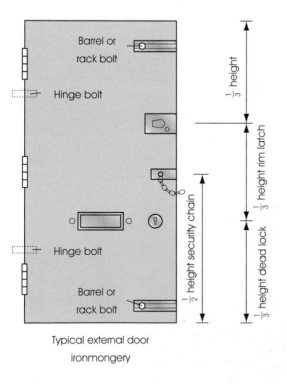

Barrel or
rack bolt

Hinge bolt

$\frac{1}{3}$ height

$\frac{1}{3}$ height rim latch

Hinge bolt

$\frac{1}{2}$ height security chain

$\frac{1}{3}$ height dead lock

Barrel or
rack bolt

Typical external door
ironmongery

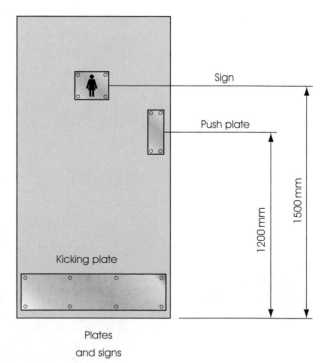

Sign

Push plate

Kicking plate

1200 mm

1500 mm

Plates
and signs

Figure 4.21 *Typical door furniture positions*

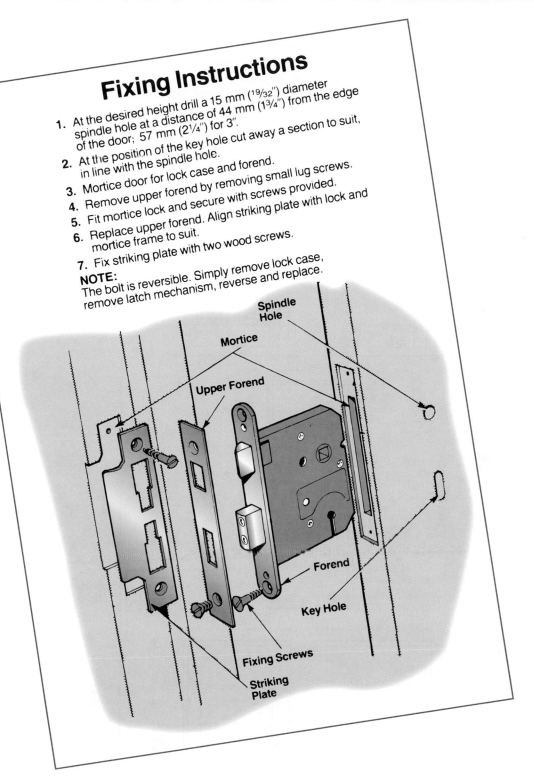

Fixing Instructions

1. At the desired height drill a 15 mm (¹⁹⁄₃₂″) diameter spindle hole at a distance of 44 mm (1¾″) from the edge of the door; 57 mm (2¼″) for 3″.

2. At the position of the key hole cut away a section to suit, in line with the spindle hole.

3. Mortice door for lock case and forend.

4. Remove upper forend by removing small lug screws.

5. Fit mortice lock and secure with screws provided.

6. Replace upper forend. Align striking plate with lock and mortice frame to suit.

7. Fix striking plate with two wood screws.

NOTE:
The bolt is reversible. Simply remove lock case, remove latch mechanism, reverse and replace.

Spindle Hole

Mortice

Upper Forend

Forend

Key Hole

Fixing Screws

Striking Plate

Second Fixing

Chapter 4

Figure 4.22 *Typical manufacturer's fixing instructions*

Before starting work, always read the job specification carefully as exact furniture positions may be stated. Also read both the door manufacturer's instructions and the ironmongery manufacturer's instructions to ensure the intended position is suitable to receive the item, e.g. the position of the lock block on a flush door, and the correct fixing procedure for a lock, etc. See Figure 4.22.

Door and ironmongery schedules

Schedules are used to record repetitive design information. Read with a range drawing and floor plans, they may be used to identify a type of door, its size, the number required, the door opening in which it fits, the hinges it will swing on and details of other furniture to be fitted to it. See Figures 4.23 to 4.25.

Figure 4.23 *Door range drawing*

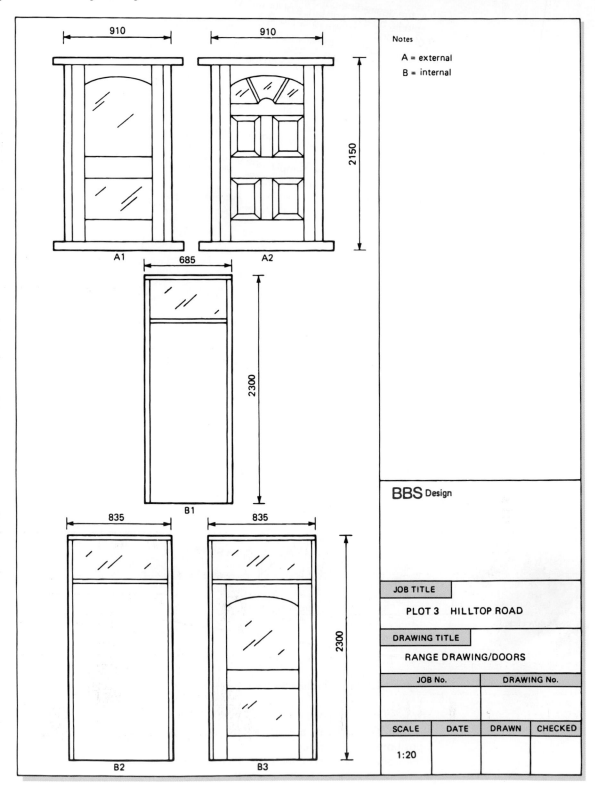

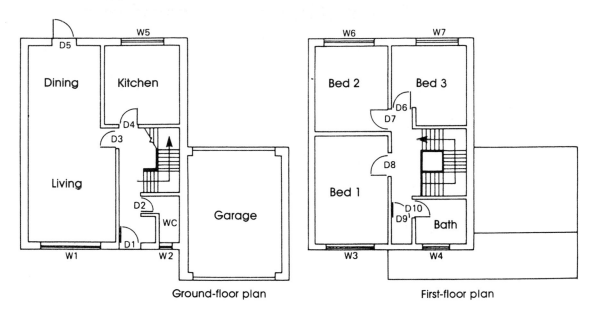

Figure 4.24 *Floor plans*

Description	D1	D2	D3	D4	D5	D6	D7	D8	D9	D10			NOTES
Type (see range)													
External glazed A1					●								
External panelled A2	●												
Internal flush B1									●				
Internal flush B2		●				●	●	●		●			
Internal glazed B3			●	●									
Size													
813 mm × 2032 mm × 44 mm	●				●								
762 mm × 1981 mm × 35 mm		●	●	●		●	●	●		●			
610 mm × 1981 mm × 35 mm									●				
Material													BBS DESIGN
Hardwood	●												
Softwood			●	●	●								
Plywood/polished		●											
plywood/painted						●	●	●	●	●			JOB TITLE PLOT 3 Hilltop Road
Infill													DRAWING TITLE Door Schedule/doors
6 mm tempered safety glass													JOB NO. DRAWING NO.
clear			●	●	●								
obscured	●												SCALE DATE DRAWN CHECKED

Figure 4.25a *Door schedule*

Description	D1	D2	D3	D4	D5	D6	D7	D8	D9	D10			NOTES
Frames													
75 mm × 100 mm (outward opening)					●								
75 mm × 100 mm (inward opening)	●												
Linings													
38 mm × 125 mm		●	●	●									
38 mm × 100 mm						●	●	●	●	●			
Shape													
Rebated stop	●				●								
Planted stop		●	●	●		●	●	●	●	●			
Transom		●	●	●		●	●	●	●	●			
Sill	●				●								
Material													BBS DESIGN
Hardwood	●												
Softwood		●	●	●	●	●	●	●	●	●			JOB TITLE — PLOT 3 Hilltop Road
Fanlight infill													DRAWING TITLE — Door Schedule/frames/lining
6 mm tempered safety glass													JOB NO. DRAWING NO.
clear													
obscured		●							●				
6 mm plywood								●					SCALE | DATE | DRAWN | CHECKED

Description	D1	D2	D3	D4	D5	D6	D7	D8	D9	D10			NOTES
Hanging													
Pair 100 mm pressed steel butt hinges			●	●	●[1.5]								
Pair 100 mm brass butt hinges	●[1.5]												
Pair 75 mm pressed steel butt hinges						●	●	●	●[1.5]	●			
Pair 75 mm brass butt hinges		●											
Fastening													
Rim night latch	●												
Mortise deadlock	●												
Mortise lock/latch		●			●					●			
Mortise latch			●	●		●	●	●	●				
100 mm brass bolts	●[2]				●[2]								BBS DESIGN
Miscellaneous													
Brass lock/latch furniture		●			●					●			
Brass latch furniture			●	●		●	●	●	●				JOB TITLE — PLOT 3 Hilltop Road
Brass letterplate	●												DRAWING TITLE
Brass knocker	●												Ironmongery schedule/doors
Brass coat hook		●[2]								●[2]			JOB NO. DRAWING NO.
Brass escutcheon	●[2]												
													SCALE | DATE | DRAWN | CHECKED

Figure 4.25b and c *Door frame/lining and door ironmongery schedules*

Details relevant to a particular door opening are indicated in the schedules by a dot or cross; a figure is also included where more than one item is required. Extracting details from a schedule is called 'taking off'. The following information concerning the WC door D2 has been taken off the schedules:

> **example**
>
> One polished plywood internal flush door type B2 762 mm × 1981 mm × 35 mm, hung on one pair of 75 mm brass butts and fitted with one mortise lock/latch, one brass mortise lock/latch furniture and two brass coat hooks.

activity

Take off the following information from the previous schedules. How many type B2 painted doors are required?

Produce a list of hinges required for the whole house.

State the size and type of door for opening D1.

Door hanging

Door hanging is normally carried out before skirtings and architraves are fixed. Speed and confidence in door hanging can be achieved by following the procedure illustrated in Figure 4.26 and outlined below:

1. Measure height and width of door opening.
2. Refer to door schedule and select correct door.
3. Locate and mark the top and hanging side of the opening and door.
4. Cut off the horns (protective extensions on the top and bottom of each stile) on panelled doors. Flush doors will probably have protective pieces of timber or plastic on each corner; these need to be prised off.
5. Where plain linings are used, tack temporary stops either side at mid height to stop the door falling through the opening. These should be set back from the edge of the lining by the thickness of the door.
6. 'Shoot' (plane to fit) in the hanging stile of door to fit the hanging side of the opening. This should be planed at a slight undercut angle to prevent binding.
7. 'Shoot' the door to width. Allow a 2 mm joint all around between door and frame or lining. Many carpenters use a 2p coin to check. The closing side will require planing to a slight angle to allow it to close. This is termed a 'leading edge'.
8. 'Shoot' the top of the door to fit the head of the opening. Saw or shoot the bottom of the door to give a 6 mm gap at floor level or to fit the threshold.
9. External doors may require rebating along the bottom edge, to fit over the water bar.
10. Mark out and cut in the hinges. Screw one leaf of each hinge to the door.
11. Offer up the door to the opening and screw the other leaf of each hinge to the frame.
12. Adjust fit as required. Remove all arrises (sharp edges) to soften the corners and provide a better surface for the subsequent paint finish. If the closing edge rubs the frame, the hinges may be proud and require the recesses being cut deeper. If the recesses are too deep, the door will not close fully and will tend to spring open, which is known as being 'hinge bound'. In this case a thin cardboard strip can be placed in the recess to pack out the hinge.

did you know?

On flush and fire check doors the hanging side should have been marked by the manufacturer. When the hanging side is not shown on the drawing, the door should open into the room to provide maximum privacy, but not onto a light switch.

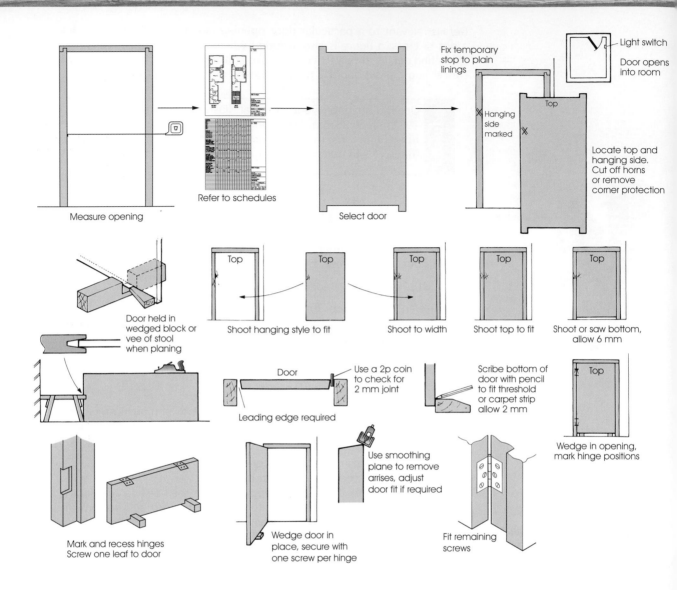

Figure 4.26 *Door hanging procedure*

13. Fit and fix the lock or latch. This correctly positions the closing edge of the door in line with the face of the frame of lining.
14. Remove temporary stops from plain linings and replace with planted stops pinned in position. Fix head first, then the sides. Square edge stops can be butted; moulded stops will require mitring. Allow a gap of up to 2 mm between the door and stop, in order to prevent binding after painting.
15. Fit any other ironmongery, e.g. bolts, letter plates, handles, etc.

It is usual practice only to fit and not to fix the other ironmongery, e.g. handles and bolts etc., at this stage. They should be fixed later during the finishing stage after all painting works are completed.

Weatherboard

Inward opening external doors in exposed locations may be fitted with a weatherboard as illustrated in Figure 4.27. These are screwed at the bottom of the external door face to throw rainwater clear of the waterbar. The ends of the weatherboard may be kept clear of the door frame or let into it.

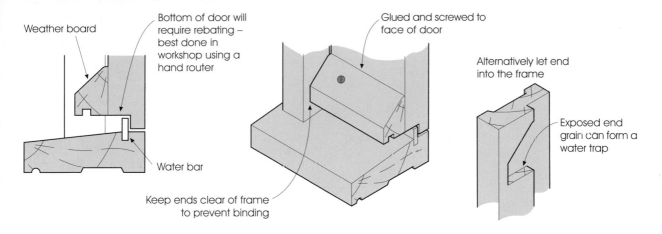

Weather board

Bottom of door will require rebating – best done in workshop using a hand router

Water bar

Keep ends clear of frame to prevent binding

Glued and screwed to face of door

Alternatively let end into the frame

Exposed end grain can form a water trap

Figure 4.27 *Weatherboard for inward-opening external doors*

Hinge recessing

The leaves of a hinge can be recessed into the door and frame equally, termed half-and-half. See Figure 4.28.

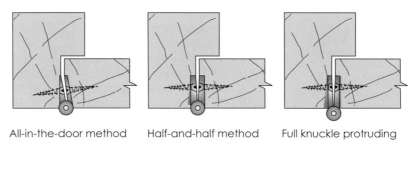

All-in-the-door method   Half-and-half method   Full knuckle protruding

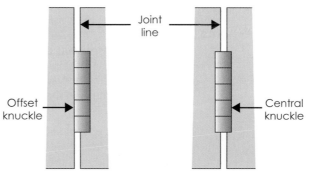

Joint line

Offset knuckle

Central knuckle

Figure 4.28 *Hinge recessing and knuckle positioning*

Alternatively the hinge can be offset, with the front edge of both leaves being recessed into the door leaving a clean, unbroken joint line. This 'all-in-the-door' method is more popular for cabinet work than door hanging. The actual method used is dependent on personal preference as each is equally as good; check with the specification, foreman or customer.

The position of the knuckle in relation to the face of the door is optional, but is sometimes dependent on the hinge. Full-size room doors are mainly hung with the full knuckle protruding beyond the door face, to give increased clearance.

Some people like to see the leaf recessed up to the centre of the knuckle for a neater appearance. However this is mainly used for cabinet doors. Once again check the specification and personal preferences.

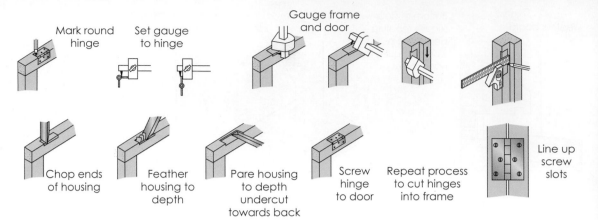

Figure 4.29 *Hinge recessing procedure*

For the correct hinge recessing procedure, follow Figure 4.29:

1. Decide on the hinge position. This will depend on the size and type of door being hung. (See 'Hinge size and positioning' on p. 110 for further details.)
2. Mark the hinge position on both the frame and the door or component to be hung.
3. Position the hinge on the edge of the door and mark the top and bottom of the leaf with a sharp pencil.
4. Repeat the process to mark the leaf position on the frame.
5. Set a marking gauge to the width of the hinge leaf and score a line on the edge of the door and frame, between the two pencil lines.
6. On frames and linings with a stuck rebated stop a combination square can be set to mark the leaf width with a pencil.
7. Reset the marking gauge to the thickness of the hinge leaf. Gauge the door face and the frame edge at each hinge position.
8. Use a chisel held vertically and mallet to chop the ends of each hinge housing position to the recess.
9. Use the chisel bevel side down and a mallet to feather each hinge housing to depth.
10. With the chisel bevel-side up pare the feathered housing to the gauged depth. Use the chisel at a slight angle so that the bottom of each recess is slightly undercut towards the back.
11. Screw the hinges to the door first. Then offer up the door and screw to the frame.

When using brass screws and hinges, it is good practice to screw the hinges on initially with a matching set of steel screws. The brass screws, being softer, are easily damaged and may even snap off when being screwed in. The steel screws pre-cut a thread into the pilot holes for the brass screws to follow easily without risk of damage. Candle wax or petroleum jelly may be applied to lubricate the screw thread before insertion. If slot-head screws are being used, these should be lined up vertically. This gives an enhanced appearance and prevents the build-up of paint or polish in the slots.

Installation of mortise dead lock, latch or lock/latch

For the installation of mortise locks and latches, follow Figure 4.30.

Always consult manufacturers' instructions.

1. Wedge the door in the open position. Use the lock as a guide to mark the mortise position on the door edge at the pre-marked height.
2. Set a marking gauge to half the thickness of the door. Score a centre line down the mortise position to mark for drilling.

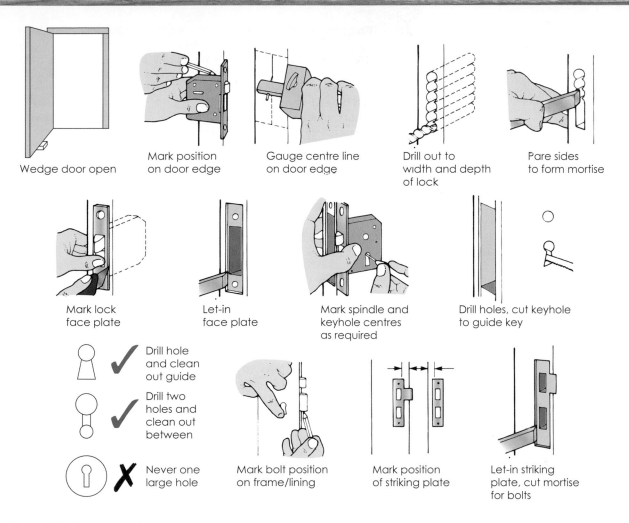

Wedge door open

Mark position
on door edge

Gauge centre line
on door edge

Drill out to
width and depth
of lock

Pare sides
to form mortise

Mark lock
face plate

Let-in
face plate

Mark spindle and
keyhole centres
as required

Drill holes, cut keyhole
to guide key

Drill hole
and clean
out guide

Drill two
holes and
clean out
between

Never one
large hole

Mark bolt position
on frame/lining

Mark position
of striking plate

Let-in striking
plate, cut mortise
for bolts

Figure 4.30 *Fitting procedure for mortise locks and latches*

Second Fixing

Chapter 4

3. Select a drill bit the same diameter as the thickness of the lock body. An oversize bit will weaken the door; use an undersize one and additional paring will be required.
4. A hand brace and bit or power drill and spade bit may be used. Mark the required drilling depth on the bit using a piece of masking tape. Drilling too deep will again weaken the door, as well as the risk in panel and glazed doors of the drill breaking out right through the stile.
5. Drill out the mortise working from the top, each hole slightly overlapping the one before.
6. Use a chisel to pare away waste between holes to form a neat rectangular mortise.
7. Slide the lock into the mortise and mark around the faceplate.
8. Remove the lock and use a marking gauge to score deeper lines along the grain, as this helps to prevent the fine edge breaking out or splitting when chiselling out the housing.
9. Use a chisel to form the housing for the faceplate (let in).
10. The surface can first be feathered (as when recessing hinges) and finally cleaned.
11. Hold the lock against the face of the door, with the faceplate flush with the door edge and lined up with the faceplate housing. Mark the centre positions of the spindle and keyhole as required with a bradawl.
12. Use a 16 mm drill bit to drill the spindle hole, working from both sides to avoid breakout, or clamp a waste piece to the back of the door.

13. Use a 10 mm drill bit to drill the keyhole, again working from both sides. Cut the key slot with a padsaw and clean out with a chisel to form a key guide. Alternatively, use a 6 mm drill bit to drill out a second hole below the 10 mm hole and chisel out waste to form a key guide. Never drill a larger hole for the key as it will not give a guide when inserting the key, making it a hit-and-miss affair.
14. Insert the lock, check the spindle and keyholes, and align from both sides. Secure the faceplate with screws. Re-check that the key works.
15. With the dead bolt out, close the door against the frame and mark the bolt and latch positions on the edge of the frame. Square these positions across the face of the frame.
16. Set the adjustable square from the face of the door to the front edge of the latch or dead bolt. Use it to mark the position on the face of the frame.
17. Hold the striking plate over the latch or dead bolt position and mark around the striking plate. Gauge vertical lines to prevent breakout when chiselling.
18. Chisel out to let in the striking plate. Again you may find it easier to feather first before finally cleaning out. The extended lead-in or lip for the latch may require a slightly deeper bevelled housing or recess.
19. Check for fit and screw the striking plate in place. Select a chisel slightly smaller than the striking plate bolt-holes.
20. Chop mortises to accommodate both the latch and dead bolt. Some striking plates have boxed bolt-holes. These must be cut beforehand.
21. Finally fit the lever furniture, knob furniture or keyhole escutcheon plates, as appropriate, and check for smooth operation. When fixing keyhole escutcheon plates, the key should be passed through the plate and into the lock and centralized on the key shaft before screwing.

Installation of a cylinder rim night latch

To fix a cylinder rim night latch, follow Figure 4.31.

Always consult manufacturers' instructions.

1. Wedge the door in an open position. Use the template supplied with the lock at the pre-marked height to mark the centre of the cylinder hole.
2. Use a 32 mm auger bit in a brace or a spade bit in a power drill to drill the cylinder hole. Drill from one side until the point just protrudes. Complete the hole from the other side to make a neat hole, avoiding breakout, or alternatively by cramping a block to the door.

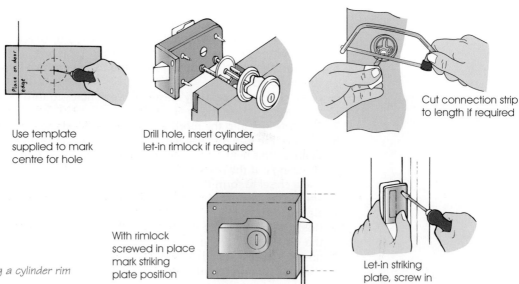

Use template supplied to mark centre for hole

Drill hole, insert cylinder, let-in rimlock if required

Cut connection strip to length if required

With rimlock screwed in place mark striking plate position

Let-in striking plate, screw in place

Figure 4.31 *Fixing a cylinder rim night latch*

3. Pass the cylinder through the hole from the outside face and secure it to the mounting plate on the inside with the connecting machine screws.
4. Ensure that the cylinder key slot is vertical before fully tightening the screws. For some thinner doors these machine screws may require shortening before use with a hacksaw. If required, secure the mounting plate to the door with woodscrews.
5. Check the projection of the flat connection strip. This operates the latch and is designed to be cut to suit the door thickness. If necessary use a hacksaw to trim the strip so that it projects about 15mm past the mounting plate.
6. Align the arrows on the backplate of the rim lock and the turnable thimble.
7. Place the rim lock case over the mounting plate, ensuring that the connection strip enters the thimble.
8. Mark out and let in the rim lock lip in the edge of the door if required.
9. Secure the rim lock case to the door or mounting plate with wood or machine screws as required. Check both the key and inside handle for smooth operation.
10. Close the door and use the rim lock case to mark the position of the keep (striking plate) on the edge of the door frame.
11. Open the door and use the keep to mark the lip recess on the face of the frame. Chisel out a recess to accommodate the keep's lip. Secure the keep to the frame using woodscrews.
12. Finally, check from both sides to ensure a smooth operation.

Installation of a letter plate

For correct installation of a letter plate, follow Figure 4.32.

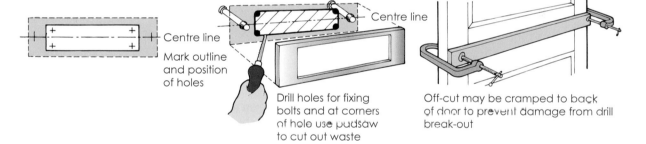

Centre line

Centre line

Mark outline
and position
of holes

Drill holes for fixing
bolts and at corners
of hole use padsaw
to cut out waste

Off-cut may be cramped to back
of door to prevent damage from drill
break-out

Figure 4.32 *Installing a letter plate*

Always consult manufacturers' instructions.

1. Wedge the door in the open position. Mark the centre line of the plate on the face of the door.
2. Position the plate over the centre line and mark around it.
3. Measure the size of the opening flap and mark the cut-out on the door. Allow about 2mm larger than the flap, to ensure ease of operation.
4. Mark the position of the holes for the securing bolts.
5. If the door is easily removed, cramp an offcut of timber to the back of the door to prevent damage from drill breakout. Alternatively the holes can be drilled from both sides.
6. Drill holes for the fixing bolts and at each corner of the flap cut-out.
7. Use a jigsaw or padsaw to saw from hole to hole.
8. Neaten up the cut-out if required using glass paper. Remove the arris from the inside edges.
9. Position the letter plate and secure, using the fixing bolts.
10. Check the flap for ease of operation and adjust if required.

Second Fixing

Chapter 4

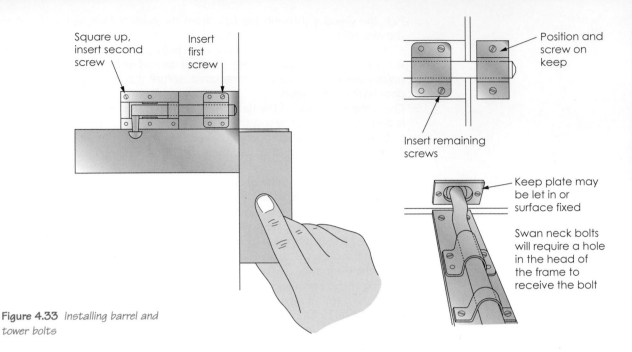

Figure 4.33 *Installing barrel and tower bolts*

Installation of barrel and tower bolts

For installation of barrel bolts, follow Figure 4.33.

Always consult manufacturers' instructions.

1. Place the bolt in the required position. Mark one of the screw holes through the backplate with a bradawl or pilot drill.
2. Insert a screw, ensuring the bolt is parallel or square to the edge of the door and insert a screw at the other end of the bolt.
3. Move the bolt to the locked position and slide the keep over the bolt.
4. Mark the screw holes in the keep and screw in place.
5. Check the bolt works smoothly before inserting the remaining screws.

Cranked or swan-necked bolts will require a hole to be drilled in the head of the door frame to receive the bolt:

1. Position and secure the bolt as before.
2. Slide the bolt to the locked position and mark around the bolt.
3. Use an auger bit slightly larger than the width of the bolt to drill a hole in the marked position. Ensure the bolt can slide to its full length.
4. Check for smooth operation.

Installation of a mortise rack bolt

For installation of a mortise rack bolt, follow Figure 4.34.

Always consult manufacturers' instructions.

1. Use a marking gauge to score a centre line on the edge of the door.
2. Mark the centre of the bolt-hole using a try square and pencil. Transfer the line onto the face of the door that the bolt will operate from.
3. Select an auger or spade bit slightly larger than the bolt barrel. Drill a hole in the edge of the door. Use a piece of tape wrapped around the bit as a depth guide.
4. Insert the bolt, and then turn the faceplate so that it is parallel to the door edge. Mark around the faceplate with a pencil.
5. Remove the bolt and gauge the parallel edges and using a sharp chisel 'let in' the faceplate.

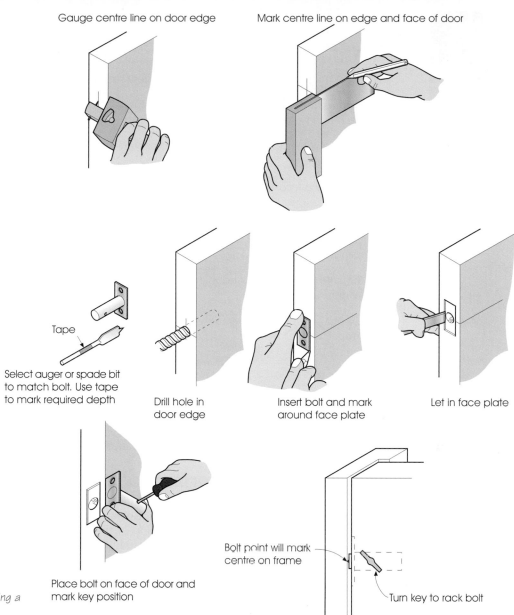

Gauge centre line on door edge

Mark centre line on edge and face of door

Tape

Select auger or spade bit to match bolt. Use tape to mark required depth

Drill hole in door edge

Insert bolt and mark around face plate

Let in face plate

Place bolt on face of door and mark key position

Bolt point will mark centre on frame

Turn key to rack bolt

Figure 4.34 *Installing a mortise rack bolt*

6. Place the bolt on the face of the door. Use the bradawl to mark the key position.
7. Use a 10 mm bit to drill the keyhole through the face of the door, into the bolthole.
8. Insert the bolt into the hole, then check that the faceplate finishes flush and that the keyhole lines up. Insert the fixing screws.
9. Insert the key through the keyhole plate into the bolt. Position the keyhole plate centrally over the key and insert the fixing screws.
10. Close the door and turn the key to rack the bolt to its locked position. The point on the bolt will mark the clearance hole centre position on the frame.
11. Drill the clearance hole for the bolt in the frame on the marked centre.
12. Close the door. Rack the bolt and check for the correct alignment of the bolt clearance hole.
13. 'Let in' the keep centrally over the clearance hole and insert the fixing screws.
14. Check for smooth operation.

1. State the purpose of door and ironmongery schedules.

2. Produce a sketch to show the typical positions of the hinges and mortise lock/latch to an external plywood flush door.

3. State why manufacturers' instructions should be followed when fitting ironmongery.

4. State the purpose of a leading edge to door stiles.

5. Define 'arris' and state why they should be removed from door edges.

6. Explain the treatment required to the top edge of a door head when hung using rising butts.

7. State why the stile of a door should be measured before ordering any ironmongery.

8. List the sequence of operations for hanging an external front door to a domestic property.

Finishing trim

Vertical and horizontal mouldings

These are often referred to as trim and include architraves, skirting, dado rail, picture rail and cornice. They are used to cover the joint between adjacent surfaces, such as wall and floor/ceiling or the joint between plaster and frames. In addition they provide a decorative feature, and may also serve to protect the wall surface from knocks, as illustrated in Figure 4.35.

Mouldings terminology

Architrave – the decorative trim that is placed internally around door and window openings to mask the joint between wall and timber and conceal any subsequent shrinkage and expansion.

Skirting – the horizontal trim, often a timber board, which is fixed around the base of a wall to mask the joint between the wall and floor. It also protects the plaster surface from knocks at low level.

Dado rail – a moulding applied to the lower part of interior walls at about waist height approximately 1 m from the floor. It is also known as a chair rail as it was designed to coincide with a tall chair back height to protect the plaster.

Picture rail – a moulding applied to the upper part of interior walls between 1.8 m and 2.1 m from the floor. Special clips are hooked over the rail in order to suspend the picture frames.

Cornice – the moulding used internally at the wall/ceiling junction. It is normally formed from plaster, only rarely from timber.

Trim – the collective term for vertical and horizontal mouldings (Figure 4.36). They are normally ready to assemble, machined to a range of standard profiles (shapes) as shown in Figure 4.37.

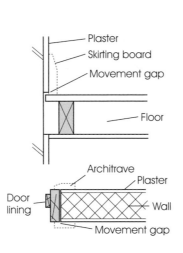

Plaster
Skirting board
Movement gap
Floor

Architrave
Plaster
Door lining
Wall
Movement gap

Skirting

Dado rail

Figure 4.35 *Covering of gaps and protecting plasterwork*

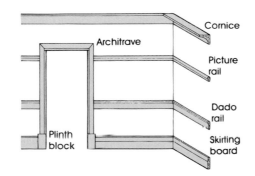

Cornice
Picture rail
Architrave
Dado rail
Plinth block
Skirting board

Figure 4.36 *Types of trim*

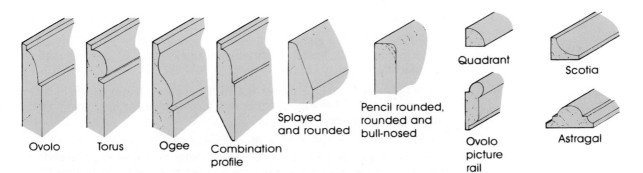

Ovolo

Torus

Ogee

Combination profile

Splayed and rounded

Pencil rounded, rounded and bull-nosed

Quadrant

Ovolo picture rail

Scotia

Astragal

Figure 4.37 *Standard trim sections*

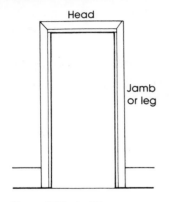

Figure 4.38 *Architraves*

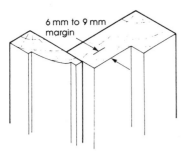

Figure 4.39 *Margin to architraves*

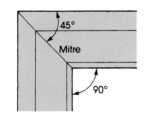

Figure 4.40 *Mitre to architraves*

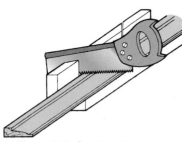

Figure 4.41 *Cutting mitre*

Skirting is often mass produced using a combination profile, e.g. with an ovolo mould on one face and edge and splayed and rounded on the other. This enables it to be used for either purpose and also reduces the timber merchant's stock range. In addition the moulding profile has the effect of an undercut edge enabling it to fit snugly to the floor surface.

Cutting and fixing trim

Solid timber, either softwood or hardwood, is used for the majority of mouldings. However, MDF (medium density fibreboard) and low density foamed core plastics are used to a limited extent for moulding production.

The following cutting and fixing details are generally suitable for all three of these materials. Consult the manufacturer's instructions prior to starting to fix other proprietary mouldings/trim.

Architraves

Figure 4.38 shows that a set of architraves consist of a horizontal head and two vertical jambs or legs.

A 6mm to 9mm margin is normally left between the frame or lining edge and the architrave (Figure 4.39). This margin provides a neat appearance to an opening; an unsightly joint line would result if architraves were to be kept flush with the edge of the opening.

The return corners of a set of architraves are mitred. For right-angled returns (90°) the mitre will be 45° (half the total angle) and can be cut using a mitre box or block as shown in Figures 4.40 and 4.41.

Mitres for corners other than right angles will be half the angle of intersection. They can be practically found by marking the outline of the intersecting trim on the frame/lining or wall, and joining the inside and outside corners to give the mitre line (see Figure 4.42). Moulding can be marked directly from this or alternatively an adjustable bevel can be set up for use.

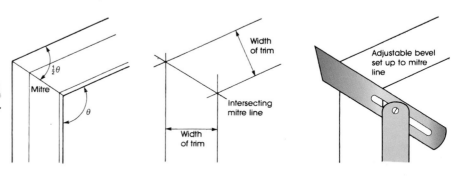

Figure 4.42 *Determining mitre for corners other than right angles*

The head is normally marked, cut and temporarily fixed in position first, as shown in Figure 4.43. The jambs can then be marked, cut, eased if required and subsequently fixed.

Where the corner is not square or you have been less than accurate in cutting the mitre, it will require easing, either with a block plane or by running a tenon saw through the mitre.

Fixing is normally direct to the door frame/lining at between 200mm and 300mm centres using typically 38mm or 50mm long oval or lost-head nails (Figure 4.44). These should be positioned in the fillets or quirks (flat surface or groove in moulding, see Figure 4.45) and punched in.

Mark margin

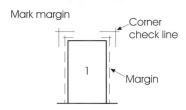

Cut first mitre, mark second

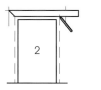

Cut second mitre and fix head

Mark mitre to first jamb

Cut mitre, fix first jamb

Mark mitre to second jamb

Cut mitre, fix second jamb

Punch in nails and remove arrises

Figure 4.43 *Marking and fixing architraves*

Mitres should be nailed through their top edge to reinforce the joint and ensure both faces are kept flush (see Figure 4.46). Suitable for this purpose are 38 mm oval or lost-head nails.

In addition, architraves, especially very wide ones, are often fixed back to the wall surface using either cut or masonry nails. (Do not forget eye protection; see Figure 4.47.)

A plinth block is a block of timber traditionally fixed at the base of an architrave to take the knocks and abrasions at floor level (see Figure 4.48). It is also used to ease fixing problems that occur when skirtings are thicker than the architrave.

In current practice plinth blocks will rarely be found except in restoration work, new high-quality work in traditional style or where the skirting is thicker than the architrave.

Architraves may be butt jointed to the plinth block, but traditionally they were joined using bare-faced tenon and screws.

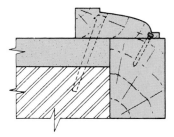

Figure 4.44 *Normal method of fixing architraves*

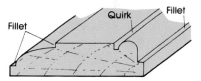

Figure 4.45 *Fixings best positioned in fillets or quirks for concealment*

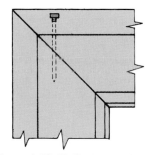

Figure 4.46 *Nailing mitre joints at corners of architraves*

Figure 4.47 *Eye protection*

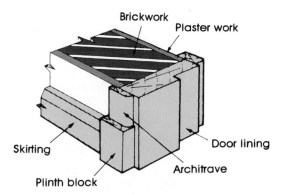

Figure 4.48 *Use of plinth blocks*

Second Fixing

Chapter 4

Edges of mouldings may be undercut to ensure a snug fit to wall and floor surfaces.

Biscuits are compressed beech ovals that fit into matching saw slots and are used to reinforce butt joints.

Suitable eye protection must be worn when driving masonry nails as these are often brittle and likely to shatter and fly off in all directions.

Corner blocks

Sometimes used at the return corners of a set of architraves in place of mitres (Figure 4.49). These may be simply pinned in place and the architrave butt jointed to them. Alternately they may be prefixed to the architrave using dowels or a biscuit. To mask the effect of subsequent block shrinkage the head and leg architraves may be housed into them.

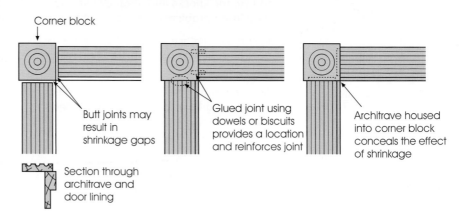

Figure 4.49 *Use of corner blocks*

Scribing architrave

Architraves should be scribed (one member cut to fit over the contour of another) to fit the wall surface, where frames/linings abut a wall at right angles.

◆ Temporarily fix the architrave jamb in position, keeping the overhang the same all the way down. Figure 4.50 shows how to set a compass to the required margin plus the overhang, or alternatively use a piece of timber this size (gauge block).
◆ Mark with the compass or gauge slip the line to be cut.
◆ By slightly undercutting the edge (making it less than 90°), it will fit snugly to the wall contour as shown in Figure 4.51.

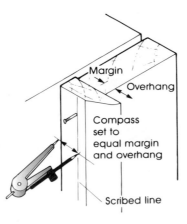

Figure 4.50 *Scribing architraves*

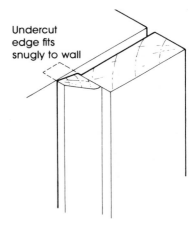

Figure 4.51 *Scribe should fit snugly to wall surface*

A quadrant mould or scotia mould is often used to cover the joint to provide a neat finish to the reveal of external door frames. Quadrant moulds may also be used in place of an architrave jamb where the frame/lining joins to a wall at right angles. The sharp arris on the quadrant mould is best pared off with a chisel, as shown in Figure 4.52, to enable the mould to sit snugly into the plaster/timber intersection.

All nails used for fixings should be punched below the surface on completion of the work. This is in preparation for subsequent filling by the painter.

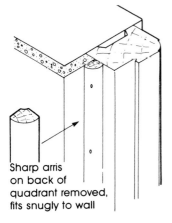

Sharp arris on back of quadrant removed, fits snugly to wall

Figure 4.52 *Use of quadrants as an alternative to scribing*

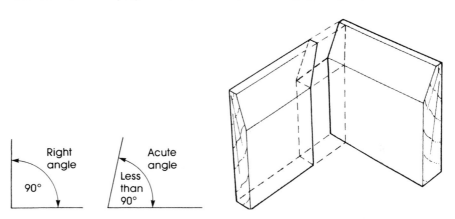

Figure 4.53 *Internal corners 90° or less*

Figure 4.54 *Scribing internal corners*

Skirting

Skirting is normally cut and fixed directly after the architrave.

Internal corners of 90° and less (right angles and acute angles) are scribed, one piece being cut to fit over the other (see Figures 4.53 and 4.54).

Scribes can be formed in one of two ways:

Mitre and scribe – fix one piece and cut an internal mitre on the other piece to bring out the profile as shown in Figure 4.55. Cut the profile square on mitre line to remove waste. Use a coping saw for the curve.

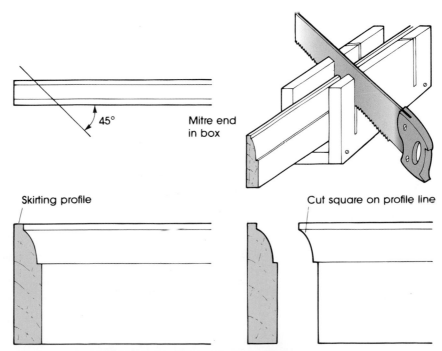

45°

Mitre end in box

Skirting profile

Cut square on profile line

Figure 4.55 *Cutting an internal scribe (mitre and scribe)*

Second Fixing

Chapter 4

Compass scribe – fix one piece; place the other piece in position. Scribe with a compass as shown in Figure 4.56. Cut square on the scribed line to remove waste.

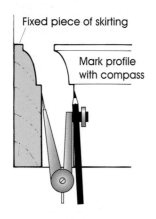
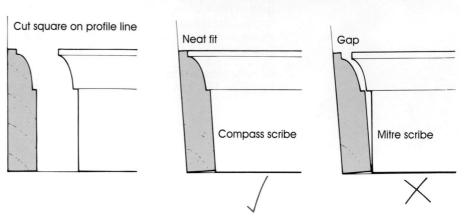

Fixed piece of skirting

Mark profile with compass

Cut square on profile line

Neat fit

Compass scribe

Gap

Mitre scribe

Figure 4.56 *Compass marking and cutting a scribe*

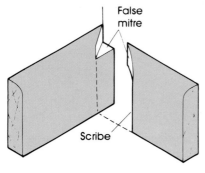

False mitre

Scribe

Figure 4.57 *False mitre and scribe*

Scribing is the preferred method, especially where the walls are slightly out of plumb. The mitred scribe would have a gap, but the compass scribe would fit the profile neatly.

Mitres are not normally used for internal corners of skirting. Wall corners are rarely perfectly square, making the fitting difficult. In addition mitres open up as a result of shrinkage, forming a much larger gap than scribes.

However, internal corners on bullnosed or pencil-rounded skirtings may be cut with a false or partial mitre on the top rounded edge and the remaining flat surface scribed to fit as shown in Figure 4.57.

Internal corners over 90°, called obtuse angles (Figure 4.58), are best jointed with a mitre.

External corners should be mitred (Figure 4.59). These return the moulding profile at a corner rather than a butt joint, as seen in Figure 4.60, which would show unsightly end grain.

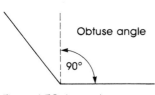

Obtuse angle

90°

Figure 4.58 *Internal corner over 90°*

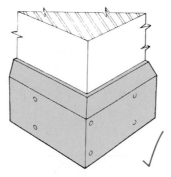

Figure 4.59 *Mitring external corners*

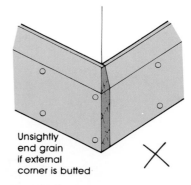

Unsightly end grain if external corner is butted

Figure 4.60 *Butting of external corners*

Mitres for 90° external corners may be cut in a mitre box.

Mitres for both internal and external corners over 90° can be marked out using the following method, illustrated in Figures 4.61 to 4.63, and then cut freehand.

1. Use a piece of skirting to mark line and width of skirting on the floor either side of the mitre (Figure 4.61).
2. Place length of skirting in position.
3. Mark position of plaster arris on the top edge of skirting. For internal corners this will be the actual back edge of the skirting (Figures 4.62 and 4.63).
4. Mark outer section on front face of skirting.
5. Use a try square to mark line across face and back surface of skirting.
6. Cut mitre freehand.
7. Repeat marking out and cutting procedure for other piece.
8. Fix skirting to wall; external corners should be nailed through the mitre.

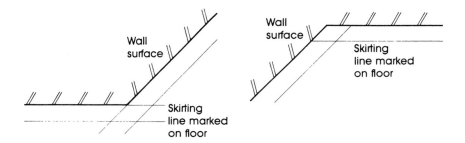

Figure 4.61 *Line of skirting for corners over 90°*

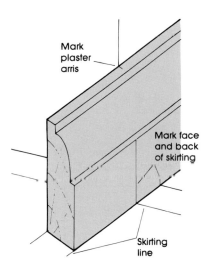

Figure 4.62 *Marking out external corners over 90°*

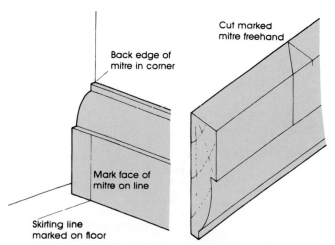

Figure 4.63 *Marking out internal corners over 90°*

Second Fixing

Chapter 4

Long lengths are fixed first as indicated in Figure 4.64, starting with those having two trapped ends (both ends between walls). Marking and jointing internal corners is much easier when one end is free.

Where the second piece to be fixed also has two trapped ends, a piece slightly longer than the actual length required, by say 50 mm, can be angled across the room or allowed to run through the door opening for scribing the internal joint (Figures 4.65 and 4.66).

After scribing and cutting to length it can be fixed in position.

Very short lengths of skirting returned around projections may be fixed before the main lengths. The two short returns are mitred at their external ends, cut square at their internal ends and fixed in position as shown in Figure 4.67.

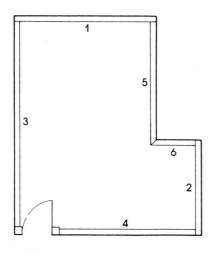

Figure 4.64 *Order of fixing (trapped ends first)*

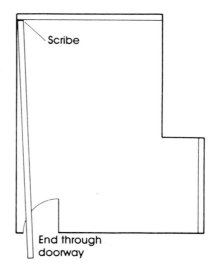

Figure 4.65 *Extend through doorway to permit scribing of joint*

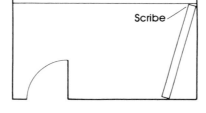

Figure 4.66 *Angle and scribe when second piece has both ends trapped*

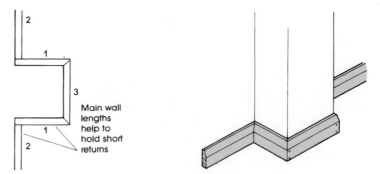

Figure 4.67 *Small pieces may be fitted first*

Main wall lengths are scribed and fixed in position. These help to hold the short returns. Finally the front piece is cut and fixed in position by nailing through the mitres.

Heading joints can be used where sufficiently long lengths of skirting are not available. Mitres are preferred to butts, because the two surfaces are held flush together by nailing through the mitre (see Figure 4.68). In addition mitres mask any gap appearing as a result of shrinkage.

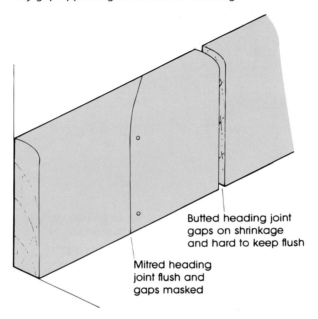

Butted heading joint gaps on shrinkage and hard to keep flush

Mitred heading joint flush and gaps masked

Figure 4.68 *Mitres are preferred for heading joints*

Where skirtings and other trim are to be fixed around a curved surface, the back face will almost certainly require kerfing. This involves putting saw cuts in the back face at regular intervals to effectively reduce the thickness of the trim. This is shown in Figures 4.69 and 4.70. The kerfs, which may be cut with a tenon saw, should be spaced between 25 mm and 50 mm apart. The tighter the curve, the closer together the kerfs should be. The depth of the kerfs must be kept the same. They should extend through the section to the maximum extent, but just keeping back from the face and top edge.

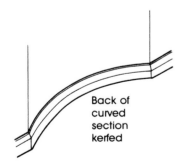

Back of curved section kerfed

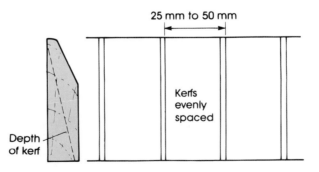

25 mm to 50 mm

Kerfs evenly spaced

Depth of kerf

Figure 4.69 *Fixing to a curved surface*

Figure 4.70 *Kerfing to the back of a skirting board*

Care is required when fixing trim back to the wall, to ensure that it is bent gradually and evenly. It is at this time that accuracy in cutting the kerfs evenly spaced to a constant depth is rewarded. Any overcutting and the trim is likely to snap at that point.

Mitres at the ends of curved sections may be marked out and cut using the same methods as described for other obtuse angles.

Skirtings and other mouldings may occasionally be required to stop part-way along a wall, rather than finish into a corner or another moulding. In these circumstances the profile should be returned to the wall or floor. This can be achieved by mitring the end and inserting a short mitred return piece or, alternatively, the return profile can be cut across the end grain of the main piece as shown in Figure 4.71.

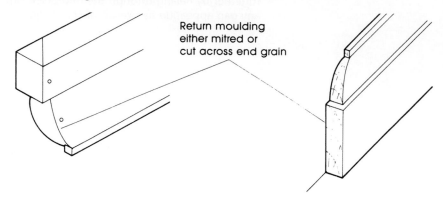

Figure 4.71 *Return the profile of moulding that stops way along a wall*

Fixing – skirting can be fixed back to walls with the aid of grounds, timber twisted plugs or direct to the surface.

Grounds – these are timber battens fixed to the wall surface using either cut nails in mortar joints or masonry nails (see Figure 4.72). One ground is required for skirtings up to 100 mm in depth. Deeper skirtings require either the addition of vertical soldier grounds at 400 mm to 600 mm centres or an extra horizontal ground. The top ground should be fixed about 10 mm below the top edge of the skirting.

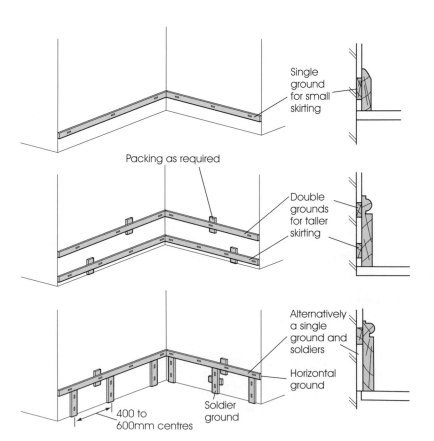

Figure 4.72 *Timber grounds for fixing skirting*

Packing pieces may be required behind the grounds to provide a true surface on which to fix the skirting. Check the line of ground with either a straight edge or string line. Skirtings can be fixed back to grounds using, typically, 38 mm or 50 mm oval or lost-head nails.

Twisted timber plugs – these are rarely used. They are shaped as shown in Figure 4.73 to tighten when driven into the raked-out vertical brickwork joints at approximately 600 mm apart. When all the plugs have been fixed, they should be cut off to provide a true line. An allowance should be made for the thickness of the plaster.

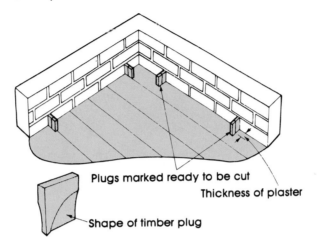

Figure 4.73 *Fixing skirting to timber plugs*

Skirtings can be fixed back into the end grain of the plugs using, typically, 50 mm cut nails. These hold better in the end grain than would oval or lost-head nails.

Prior to fixing mouldings across any wall, a check should be made to see if any services are hidden below the wall surface. Fixing into electric cables and gas or water pipes is potentially dangerous and expensive to repair. Wires to power points normally run vertically up from the floor. Wires to light switches normally run vertically down from the ceiling. Therefore, keep clear of these areas when fixing. Buried pipes in walls are harder to spot. Vertical pipes may just be seen at floor level; outlet points may also be visible (see Figure 4.74). Assume both of these run the full height of the wall, both up and down. Once again, keep clear of these areas when fixing. When in doubt an electronic device can be used to scan the wall surface prior to fixing. This gives off a loud noise when passing over buried pipes and electric cables.

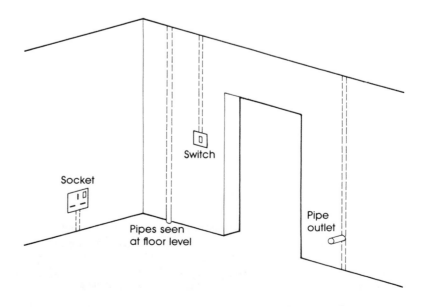

Figure 4.74 *Locating hidden services*

Second Fixing

Chapter 4

Direct to the wall – skirtings are fixed back to the wall after plastering, using typically either 50 mm cut nails or 50 mm masonry nails depending on the hardness of the wall. Oval or lost-head nails may be used to fix skirtings to timber studwork partitions.

Figure 4.75 shows that fixings should be spaced at 400 mm to 600 mm centres. These should be double nailed near the top and bottom edge of the skirting or alternatively they may be staggered between the top and bottom edge. Remember all nails should be punched below the surface.

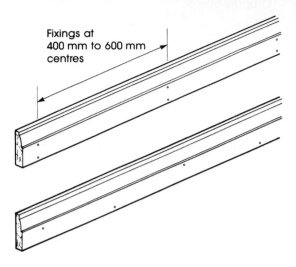

Fixings at 400 mm to 600 mm centres

Figure 4.75 *Spacing of fixings*

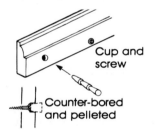

Cup and screw

Counter-bored and pelleted

Figure 4.76 *Screws sometimes used to fix hardwood skirtings*

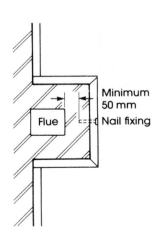

Minimum 50 mm

Flue

Nail fixing

Figure 4.78 *Fixings near flues*

Hardwood skirtings for very high quality work may be screwed in position. These should be counterbored and filled with cross-grained pellets on completion, or brass screws and cups as shown in Figure 4.76 should be used.

As a modern alternative to nails and screws, skirting and other trim fixed to wall surfaces may be bonded using a gun-applied gap-filling 'no-nails' adhesive. One or two continuous 6 mm beads should be applied to the back of the trim before positioning and pressing in place. Strutting or temporary nails (Figure 4.77) may be required to hold the trim in place while the adhesive cures. These should be left overnight before removal.

Temporary strutting

'No nails' gap-filling adhesive

Temporary nail

Figure 4.77 *Use of gap-filling adhesive to fix skirting*

The Building Regulations restricts the use of combustible materials (including timber) around heat-producing appliances and their flues. In general no structural timber (joists and rafters) is to be built into a flue, or be within 200 mm of the flue lining, or nearer than 40 mm to the outer surface of a flue.

Skirtings, architraves, mantel shelves and other trim are non-structural and therefore exempt from this requirement. However, any metal fixings associated with these must be at least 50 mm from the flue (see Figure 4.78). Metal rapidly conducts heat, which could cause the trim to catch fire. Therefore, do not use overlong fixings in this situation.

Figure 4.79 *Keeping skirting tight to floor surface when fixing*

When fixing narrow skirtings, say 75 mm to 100 mm in depth, they may be kept tight down against a fairly level floor surface with the aid of a kneeler, a short piece of board placed on the top edge of the skirting and held firmly by kneeling on it as shown in Figure 4.79.

Deeper skirtings and/or uneven floor surfaces may require scribing to close the gaps before fixing. This is carried out if required, after jointing but prior to fixing, as follows.

1. Place cut length of skirting in position.
2. Use gauge slip or compass set to widest gap to mark on the skirting a line parallel to the uneven surface as shown in Figure 4.80.

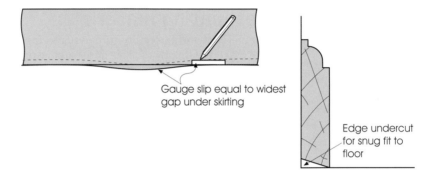

Gauge slip equal to widest gap under skirting

Edge undercut for snug fit to floor

Figure 4.80 *Scribing and cutting skirting to an uneven floor surface*

3. Trim skirting to line using either a handsaw or plane.
4. Undercut the back edge to ensure the front edge snugly fits the floor contour.

Dado and picture rails

Dado and picture rails, being horizontal mouldings, may be cut and fixed using similar methods to those used for skirtings.

Start by marking a level line in the required position around the walls as shown in Figure 4.81. (Use a straight edge and spirit level or a water level and chalk line.) The required position may be related to a datum line where established.

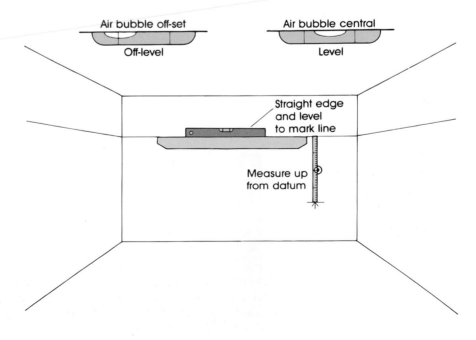

Air bubble off-set
Off-level

Air bubble central
Level

Straight edge and level to mark line

Measure up from datum

Figure 4.81 *Marking positions of dado and picture rails*

Second Fixing

Chapter 4

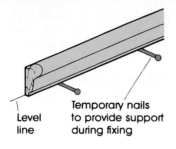

Level line

Temporary nails to provide support during fixing

Figure 4.82 *Temporary support for dado and picture rails during fixing*

When working single-handed, temporary nails can be used at intervals to provide support prior to fixing (see Figure 4.82).

Simple sections can be scribed at internal corners, as are skirtings; otherwise, use mitres. External corners should be mitred.

Fixings are normally direct to the wall surface at about 400 mm centres using typically either 50 mm cut nails or 50 mm masonry nails depending on the hardness of the wall. When fixing mouldings to timber studwork partitions, 50 mm oval or lost-heads can be used.

Mitres around external corners should be secured with nails through their edge. Typically 38 mm ovals or lost-heads are used for this purpose.

Remember that all nails should be punched below the surface.

Estimating materials

To determine the amount of trim required for any particular task is a fairly simple process, if the following procedures are used:

Architraves

The jambs or legs in most situations can be taken to be 2100 mm long. The head can be taken to be 1000 mm. These lengths assume a standard full-size door and include an allowance for mitring the ends. Thus the length of architrave required for one face of a door lining/frame is 5200 mm or 5.2 m.

Multiply this figure by the number of architrave sets to be fixed. This will determine the total metres run required, say eight sets, both sides of four door openings:

$$5.2 \times 8 = 41.6 \text{m}$$

Skirtings and other horizontal trim

Skirtings and other horizontal trim can be estimated from the perimeter. This is found by adding up the lengths of the walls in the area. The widths of any doorways and other openings are taken away to give the actual metres run required.

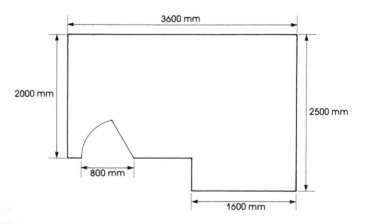

Figure 4.83 *Floor plan of room*

> **example**
>
> Determine the total length of timber required for the room shown in Figure 4.77.
>
> $$\text{Perimeter} = 2 + 3.6 + 2.5 + 1.6 + 0.5 + 2$$
> $$= 12.2 \text{m}$$
>
> Total metres run required $= 12.2 - 0.8$ (door opening)
> $$= 11.4 \text{m}$$
>
> An allowance of 10% for cutting and waste is normally included in any estimate for horizontal moulding.

example

Determine the total run of skirting required in metres for the run shown including an allowance of 10% for cutting and waste.

> Total metres run required = 11.4m
>
> Total metres run required including a 10% cutting and waste allowance
>
> = 11.4 + 1.14
>
> = 12.54m, say 12.5m

activity

Determine the total metres run of skirting and dado rail required for the room shown in the diagram. Include an allowance of 10% for cutting and waste.

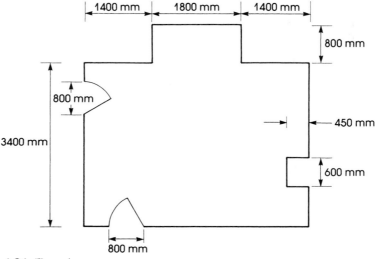

Figure 4.84 *Floor plan*

Determine the total run of architrave required in metres for both faces of the doors that open into the room shown.

measuring up

9. State the purpose of using architraves and skirtings.

10. State the reason why a margin is left between the edge of a frame and the architrave.

11. State the purpose of mitring architraves.

12. Describe a plinth block and state where it may be used.

13. Explain the reason why architraves and skirtings may be scribed.

14. Produce a sketch to show a situation where a quadrant mould has been used in the place of one architrave jamb.

15. Explain why scribes are used in preference to mitres for the internal corners of skirting.

16. Explain the situation where heading joints may be used in skirtings.

17. Describe how skirting may be prepared to be fixed around a curved surface.

18. Describe a situation where the profile of a moulding may be returned to the wall surface.

19. Produce a sketch to show a torus mould to the edge of a skirting.

20. Name the fixing used to secure an architrave to a door lining.

Encasing services

Encasing terminology

Encasing – the casing or boxing in of services. The term casing or boxing refers to the framework, cladding and trim used to form an enclosure in which service pipes are housed.

Service pipes – the system of pipes for gas, water or drainage. These are normally fixed within or on the surface of floors and walls. Service pipework is cased or boxed in to conceal the pipes, thus providing a neat, tidy appearance, which when decorated blends with the main room decoration. In addition they must also provide access to stop valves (stop cocks), drain down valves and cleaning or rodding points.

Encasing guidelines

In many situations encasing services is a simple process of forming an L-shaped or U-shaped box from timber battens, covered with a plywood or hardboard facing (see Figures 4.85 to 4.87).

Consider the following simple rules when planning and fixing casings.

◆ Use standard sections of timber where possible.
◆ Where casing is to be tiled, ensure the dimensions are simple widths of whole or half tiles.
◆ Note any stop valve or other fitting, which may require access. Fit a separate length of facing board over this section.
◆ Use WBP (weather and boil proof) plywood or an oil-tempered hardboard for casings in wet areas, e.g. kitchens, bathrooms and laundries etc.

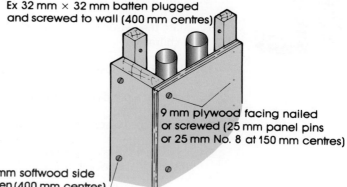

Figure 4.85 *Small corner pipe casing*

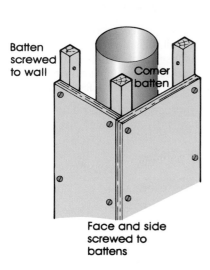

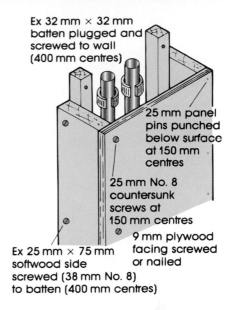

Figure 4.86 *Large corner pipe casing*

Figure 4.87 *Small pipe casing in run of wall*

Battens, typically 32 mm × 32 mm may be fixed to the wall using plugs and screws or nails, cut or masonry, depending on the wall hardness.

When using 6 mm plywood or 6 mm hardboard, timber ladder frames (see Figure 4.88) are required for support. These are typically 25 mm × 50 mm softwood half lapped together, but 9 mm and 12 mm plywood can be used for casing sides direct to battens without a supporting framework. Plywood facing can be nailed or screwed to battens/frames.

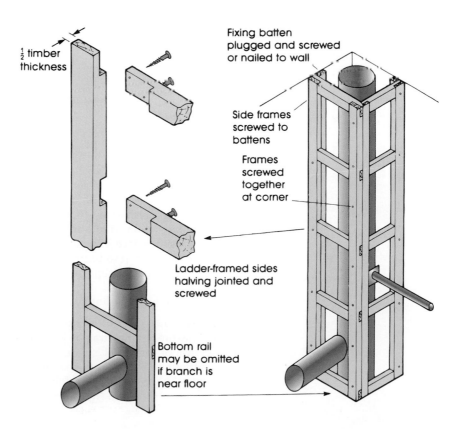

Figure 4.88 *Use of ladder frame for pipe casing*

Second Fixing **Chapter 4**

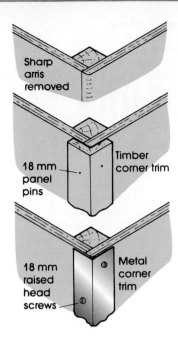

Figure 4.89 *Alternative corner treatments*

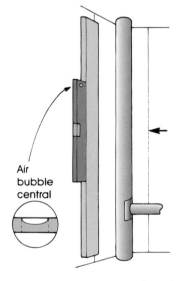

Figure 4.90 *Marking plumb line on wall*

Access panels can be screwed in position using brass cups and screws typically 25 mm No. 8. Alternatively they may be hinged as a small door.

Casings have to be scribed to wall surfaces and finished more neatly if they are to be painted or papered rather than tiled. All nails and screws should be punched or sunk below the surface ready for subsequent filling by the painter. The sharp corner arris needs to be removed with glass paper or, as an alternative, may be covered with a timber, metal or plastic trim, as shown in Figure 4.89.

Where casings are located in living rooms or bedrooms they can be packed out with fibreglass, mineral wool or polystyrene in order to quieten the noise of water passing through.

L-shaped casings

L-shaped casings are used for pipes in a corner, as follows.

◆ Mark plumb lines on walls. (Use a spirit level and straight edge, as shown in Figure 4.90.)
◆ Fix battens to the marked lines.
◆ Fix the side to the batten.
◆ Fix the front facing to the side and batten.

OR

◆ Make up and fix a ladder frame and fix the facings.

U-shaped casings

U-shaped casings are used for pipes in the middle of a wall.

◆ Mark plumb lines on the wall.
◆ Fix battens to the marked lines.
◆ Fix the sides to the battens.
◆ Fix the front to the sides.

Where pipes branch off the main one, the side will require notching or scribing over them.

For small pipes simply mark the side, drill a hole and saw the side to form a notch as shown in Figure 4.91.

Larger branch pipes are best scribed around with the face split on the pipe's centre line as shown in Figure 4.92.

Uneven wall surfaces

Sides and faces that fit to an uneven wall surface will require scribing as shown in Figures 4.93 to 4.95.

◆ Position the side against the wall, place a pencil on the wall surface and move it down to mark a parallel line on the side. Plane to the line, slightly undercutting to ensure a tight fit.
◆ Cut the ply face oversize (say 15 mm over required width).
◆ Position the face against the wall.
◆ Temporarily nail the face in position, keeping the overhang on the side the same distance from top to bottom.
◆ Set a compass or gauge slip to the width of overhang. Mark a parallel line on the face.
◆ Plane or saw to this line, slightly undercutting to ensure a tight fit.

Access panels – the edges of access panels are often chamfered as shown in Figure 4.96. This breaks the straight joint and permits a better paint finish. In addition, removal of the panel is eased without risk of damage. Simply remove screws, run a trimming knife along chamfered joints to cut paint film and lift off panel. Alternatively, access panels may be hinged into a lining and finished with an architrave trim as shown in Figure 4.97.

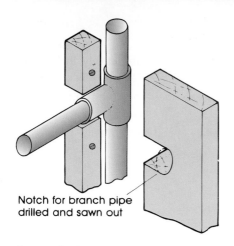

Notch for branch pipe drilled and sawn out

Figure 4.91 *Cutting around small branch pipe*

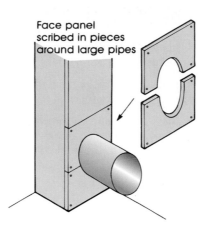

Face panel scribed in pieces around large pipes

Figure 4.92 *Cutting around large branch pipe*

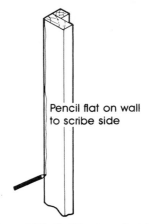

Pencil flat on wall to scribe side

Figure 4.93 *Scribing to wall with pencil*

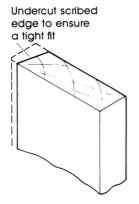

Undercut scribed edge to ensure a tight fit

Figure 4.94 *Undercutting scribes*

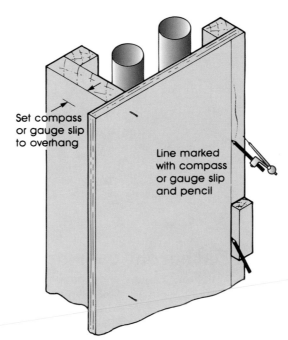

Set compass or gauge slip to overhang

Line marked with compass or gauge slip and pencil

Figure 4.95 *Scribing to wall using compass or gauge slip*

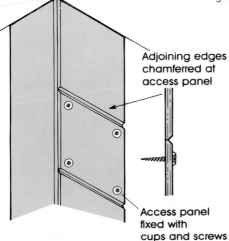

Adjoining edges chamferred at access panel

Access panel fixed with cups and screws

Figure 4.96 *Access panel with chamfered edge*

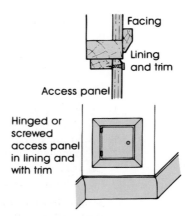

Facing

Lining and trim

Access panel

Hinged or screwed access panel in lining and with trim

Figure 4.97 *Access panel in lining*

Second Fixing

Chapter 4

Chapter 4 Second Fixing

did you know?

Filling the space behind a bath panel with fibreglass or mineral wool insulation not only provides sound insulation but also helps to keep the water in the bath hotter for longer.

Horizontal pipe casings

Horizontal pipe casings are mainly at skirting level. They can be formed using the same construction as vertical casings (Figure 4.98). Alternatively, they can be formed using a skirting board fixed to a timber top as shown in Figure 4.99.

Taller horizontal casings may have their top extended over the front facing in order to form a useful shelf (Figure 4.100).

◆ Mark a level line on the wall. (Use a spirit level and straight edge.)
◆ Mark a straight line on the floor.
◆ Fix battens to the marked lines.
◆ Fix the top and front facing.

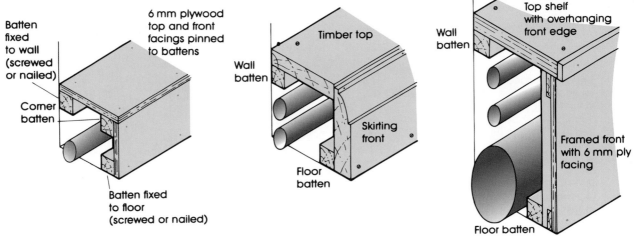

Figure 4.98 *Horizontal pipe casing*

Figure 4.99 *Skirting pipe casing*

Figure 4.100 *Shelf top to horizontal pipe casing*

Bath front casings

Bath front casings are commonly termed bath panels. They can be either a standard, normally plastic, set or purpose made.

Standard panels

Standard panels normally simply fit up under the bath rim and are fixed along their bottom edge to a floor batten as shown in Figure 4.101. Read the specific instructions supplied with the panel for details.

Purpose-made panels

Purpose-made panels have to be fixed (nailed or screwed) to a batten framework as shown in Figure 4.102. This is typically made from ex 25 mm × 50 mm PAR softwood, halved and screwed together.

The panels can be formed from a variety of materials. For example:

◆ 9 mm plywood covered with tiles;
◆ 9 mm plywood covered with carpet;
◆ 9 mm veneered plywood with applied mouldings to create a traditional panelled effect;
◆ matchboarding, TG & V (tongued, grooved and vee jointed);
◆ 3 mm melamine-faced hardboard.

An access panel is often formed at the tap end of the bath for maintenance purposes, rather than removing the whole panel. Figure 4.103 shows how this may be screwed in position or hinged on.

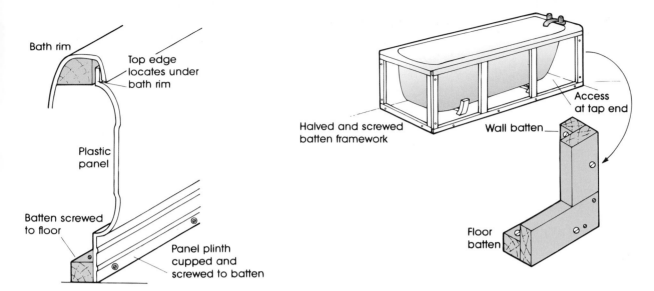

Bath rim

Top edge
locates under
bath rim

Plastic
panel

Batten screwed
to floor

Panel plinth
cupped and
screwed to batten

Figure 4.101 *Fixing standard bath panel*

Halved and screwed
batten framework

Wall batten

Access
at tap end

Floor
batten

Figure 4.102 *Batten framework for a bath panel*

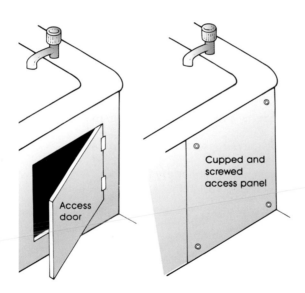

Access
door

Cupped and
screwed
access panel

Figure 4.103 *Bath access panels*

◆ Ensure the batten framework is set back far enough under the bath rim to allow for the panel thickness (see Figure 4.104).
◆ Mark plumb lines on the walls.
◆ Mark a straight line on the floor.
◆ Fix wall and floor battens to the marked lines.
◆ Make and fix the battened framework.
◆ Fix the panel to the framework.

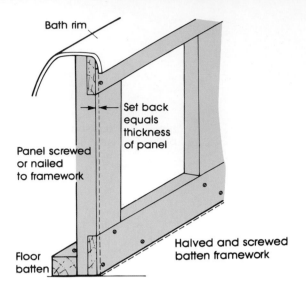

Bath rim

Set back
equals
thickness
of panel

Panel screwed
or nailed
to framework

Halved and screwed
batten framework

Floor
batten

Figure 4.104 *Positioning batten framework*

measuring up

21. State the reason why WBP plywood is used for casings in wet areas.

22. State the reason why pipe casings located in living rooms or bedrooms may be packed out with fibre glass or mineral wool.

23. State why a removable access trap may be included in pipe casings.

24. State the purpose of encasing services.

25. Explain why a supporting framework is required when using thin sheet material for pipework casings.

Kitchen units and fitments

Kitchen units fall into two distinct categories:

◆ *Purpose made* – a unit made in a joiner's shop for a specific job. Most will be fully assembled prior to their arrival on site.
◆ *Proprietary* – a unit or range of units mass produced to standard designs by a manufacturer. Budget-priced units are often sent in knock-down form (known as flat packs) ready for on-site assembly. Better quality units are often ready assembled in the factory (known as rigid units).

The two main methods of construction for both proprietary and purpose-made units, shown in Figure 4.105, are box construction and framed construction.

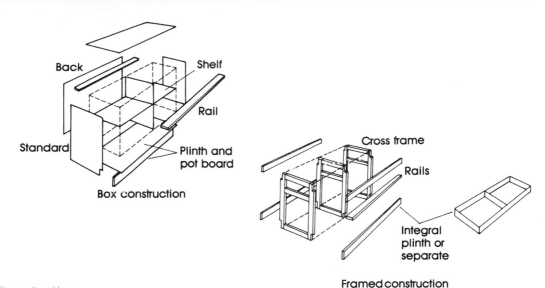

Figure 4.105 *Unit construction*

Box construction

This is also known as slab construction. This uses vertical standards and rails and horizontal shelves.

A back holds the unit square and rigid. The plinth and pot board are normally integral with the unit.

Proprietary units are almost exclusively made from 15 mm to 19 mm thick melamine-faced chipboard or medium density fibre board (MDF). Purpose-made units may be constructed using chipboard, MDF, blockboard, melamine-faced board or, more rarely, solid timber.

Flat packs use knock-down fittings or screws to join the panels. Assembly is a simple process of following the manufacturer's instructions and drawings, coupled with the ability to use a screwdriver.

Rigid and purpose-made units may be either dowelled or housed together. Glue is used on assembly to form a rigid carcass.

Framed construction

This is also known as skeleton construction. This uses frames either front and back joined by rails or standards, or cross-frames joined by rails. The plinth and pot board are normally separate items.

The frames of proprietary units are normally dowelled, whereas purpose-made units would be mortised and tenoned together.

Flat pack assembly and installation

The method of assembling and installing flat-pack units will vary from manufacturer to manufacturer. However, each unit is supplied with its own instructions. It is most important to take the time to read through these prior to commencing work. In general this is a three-stage process.

Assembly – put carcasses together. Unpack and assemble units one at a time and check contents. Open more than one and you risk confusing the parts!

Installation – fix base units to the wall starting with corner base unit and working outwards from either side. Finally install wall units, again working from the corner outwards.

Finishing – fit worktops, drawers and doors. This should not be done until all units are firmly fixed to the wall and connected together.

Typical assembly instructions are shown in Figure 4.106.

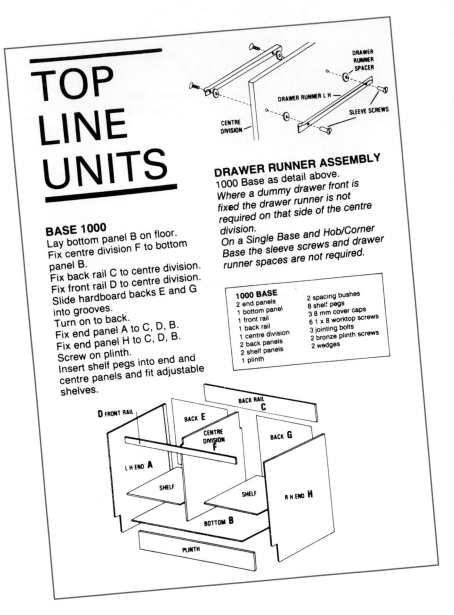

TOP LINE UNITS

BASE 1000
Lay bottom panel B on floor.
Fix centre division F to bottom panel B.
Fix back rail C to centre division.
Fix front rail D to centre division.
Slide hardboard backs E and G into grooves.
Turn on to back.
Fix end panel A to C, D, B.
Fix end panel H to C, D, B.
Screw on plinth.
Insert shelf pegs into end and centre panels and fit adjustable shelves.

DRAWER RUNNER ASSEMBLY
1000 Base as detail above.
Where a dummy drawer front is fixed the drawer runner is not required on that side of the centre division.
On a Single Base and Hob/Corner Base the sleeve screws and drawer runner spaces are not required.

1000 BASE	
2 end panels	2 spacing bushes
1 bottom panel	8 shelf pegs
1 front rail	3 8 mm cover caps
1 back rail	6 1 x 8 worktop screws
1 centre division	3 jointing bolts
2 back panels	2 bronze plinth screws
2 shelf panels	2 wedges
1 plinth	

Figure 4.106 *Typical manufacturer's assembly instructions*

Fixing base units
See Figure 4.107.

1. Position base units level and plumb using wedges provided if required. Ensure all top edges and fronts of units are flush. Secure units together using connecting screws.
2. Drill, plug and screw units to wall, using the brackets provided.
3. Trim floor wedges flush to unit with a knife.

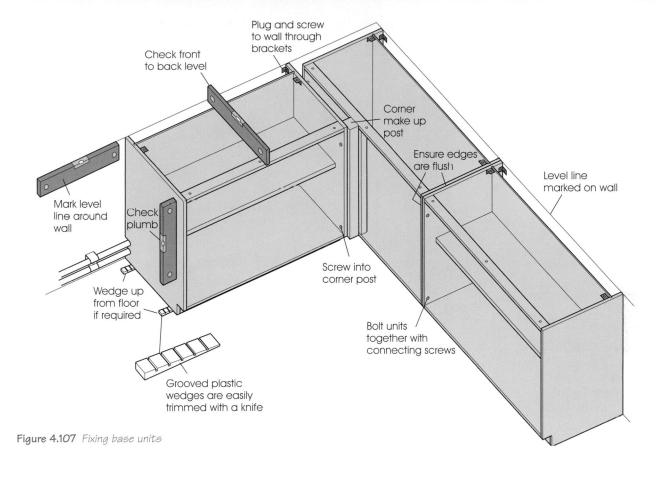

Plug and screw
to wall through
brackets

Check front
to back level

Corner
make up
post

Ensure edges
are flush

Level line
marked on wall

Mark level
line around
wall

Check
plumb

Wedge up
from floor
if required

Screw into
corner post

Bolt units
together with
connecting screws

Grooved plastic
wedges are easily
trimmed with a knife

Figure 4.107 *Fixing base units*

Fixing wall units

See Figure 4.108.

1. Draw a level horizontal line from the top of the tall unit if being used. It is normal practice to keep tall units and wall units level. Draw another line the depth of the wall units below this line. This marks the position of the underside of the wall units. Where tall units are not being used, wall units are normally fixed with a gap of 450 mm between their underside and the work surface.
2. Temporarily fix a batten on the marked level line, to act as a support while marking fixing holes and screwing.
3. Rest the unit on the batten. Ensure the tops of wall units and tall units are flush.
4. Mark the wall through the fixing holes.
5. Drill, plug and screw the unit to the wall.
6. Packing behind a fixing may be required on an uneven wall surface to ensure the units are plumb.
7. Position the remaining wall units in place one at a time.
8. Ensure the top edges and fronts are flush and secure together using connecting screws.

Second Fixing

Chapter 4

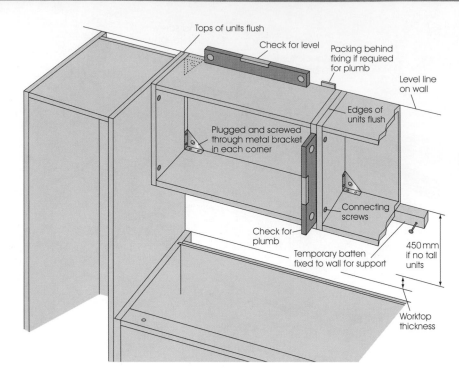

Figure 4.108 *Fixing wall units*

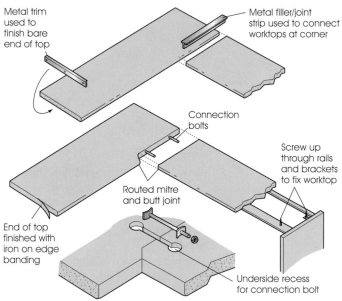

Figure 4.109 *Fixing post-formed worktops*

Fixing worktops

See Figure 4.109.

1. Measure and cut the worktop to the required size. (Post-formed, see 'Worktops', following, for other types.)
2. Metal filler/joint strips or routed butt and mitre joints are used to connect worktops that turn a corner. These joints can be pulled up tight using worktop connecting bolts. It is good practice to apply a silicone seal before pulling up the joint.
3. Position worktop and screw in place through the fixing brackets.
4. Sawn bare ends of the top can be covered with a metal trim or a plastic pre-glued edge banding. Iron edge banding into place, using a sheet of paper in between banding and iron for protection.

Fixing drawers

Insert drawers, made up previously, onto drawer runners.

Fixing doors

1. Lay the doors face down on a flat clean surface.
2. Locate the hinges over the previously fixed hinge plate and secure hinges with a mounting screw.
3. Adjust the hinges if required to ensure accurate door alignment.

Rigid and purpose-made installation

The installation of rigid and purpose-made units follows the general procedures used for flat-pack units.

Good quality rigid units have adjustable legs for easier levelling on an uneven floor. In addition wall units often have adjustable wall brackets, which enable fine adjustment to plumb and level. Decorative trim such as cornice, light pelmits and end panels may also require fitting as illustrated in Figure 4.110.

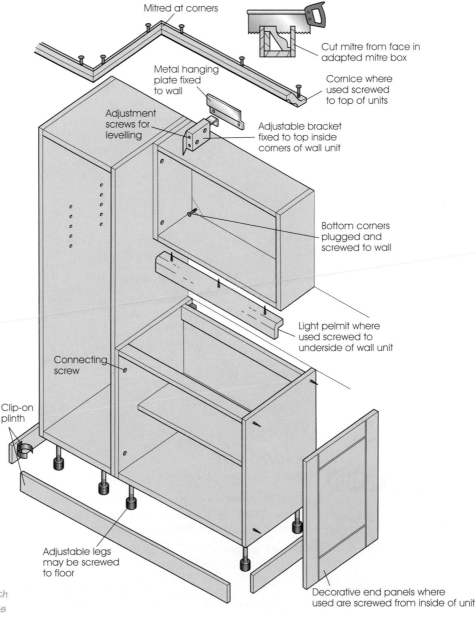

Figure 4.110 *Fixing units with adjustable legs and brackets*

Purpose-made units often have provision for scribing to uneven floor and wall surfaces (see Figure 4.111). The units should be wedged up off the ground until they are plumb and level. A compass can be set to the widest gap and used to mark a line parallel with the floor. This is then cut and the operation repeated to scribe to the wall. The unit can then be screwed to battens fixed to the wall and floor. Alternatively, cover moulds can be used in place of scribing to mask any gap between the unit and an out-of-plumb wall. The cover moulds may themselves require scribing. Plastic laminate is often cut and glued to the plinth after fixing to provide a neat, easily cleaned finish.

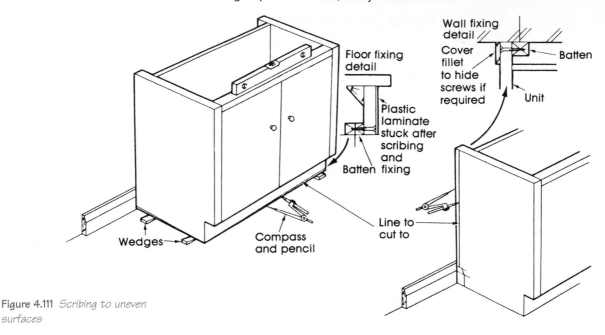

Figure 4.111 *Scribing to uneven surfaces*

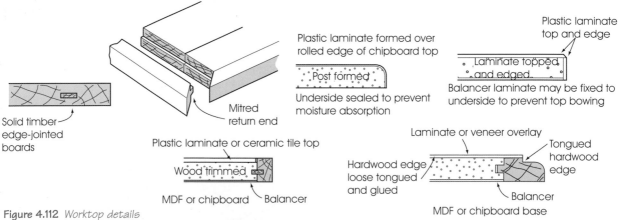

Figure 4.112 *Worktop details*

Worktops

The main types of worktop in common use are illustrated in Figure 4.112.

Post-formed – a chipboard base covered with a plastic laminate that has been formed over a rolled edge. The most popular type of worktop for proprietary units, it is ready finished and simply requires fixing in place.

Wood trimmed – a chipboard or MDF base covered with either a plastic laminate or ceramic tiles. A hardwood trim is tongued and glued to the front edge, providing a neat finish. This is mainly used for higher quality work. Hardwood trim is normally supplied loose, ready for mitring and gluing on site. An alternative is to pre-edge the baseboard and then overlay with a laminate or veneer. The overlay is trimmed off back to the edging and often incorporates a decorative mould.

Laminate topped and edged – a chipboard or MDF base edged and covered in plastic laminate, rarely used for standard jobs, as post-formed worktops are readily available at a low cost.

26. Describe the difference between flat-pack and rigid proprietary kitchen units.

27. Produce a sketch to show the following THREE worktop edge details:
    (a) post-formed
    (b) wood trimmed
    (c) laminate edged.

28. Describe a method of jointing post-formed worktops where they return around a corner.

29. A kitchen is to be fitted with a range of units to form an L shape. Describe the sequence in which these units should be fixed.

Panelling and cladding

Panelling

Wall panelling is the general term given to the covering of internal wall surfaces and sometimes ceilings, with timber or other materials to create a decorative finish (Figure 4.113). All panelling may be classified by its height and method of construction.

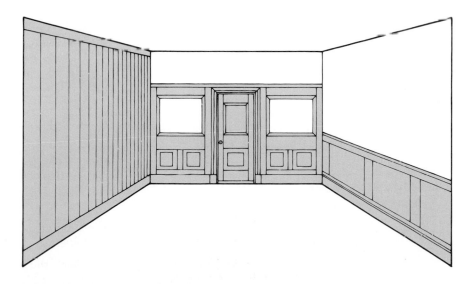

Figure 4.113 *Panelling*

Heights of panelling (see Figure 4.114)

Dado height panelling – extends from the floor up the walls to the window sill level or chair back height, i.e. about 1 m above the floor.

Three-quarter height panelling – also known as frieze height panelling. It extends from the floor up the walls to the top of the door, i.e. about 2 m.

did you know?

The procedure for applying plastic laminate is fully covered in the level one companion book, *Wood Occupations*.

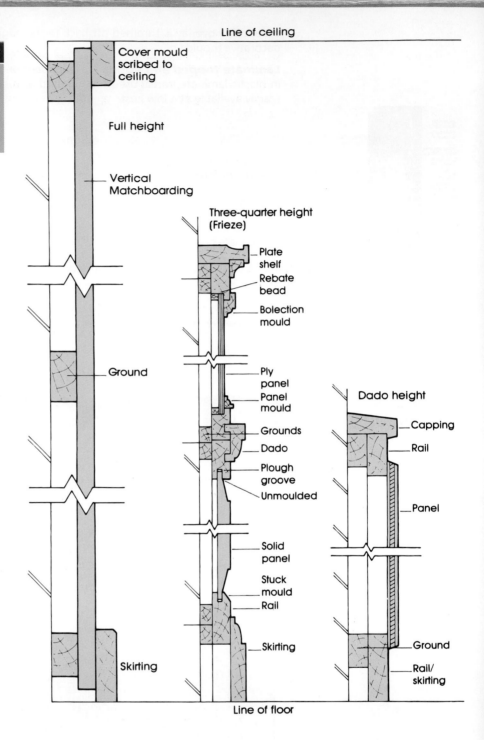

Figure 4.114 *Panelling heights*

Traditionally, a plate shelf was incorporated on top of this type of panelling to display plates and other frieze ornaments.

Full height panelling – as its name suggests, covers the whole of the wall from floor to ceiling.

Construction of panelling

Traditional panelling consists of stiles, rails and muntins, mortised and tenoned together, with panels infilled between the framing members, which are themselves fixed back to grounds. In modern usage the term panelling is also loosely applied to wall linings made up of sheet material, or matchboarding.

Grounds

Grounds provide a flat and level surface on which the panelling can be fixed (see Figure 4.115). They are normally preservative-treated softwood and may have been framed up using halvings or mortise-and-tenon joints or, alternatively, supplied in lengths for use as separate grounds or counter battening (a double layer fixed at right angles).

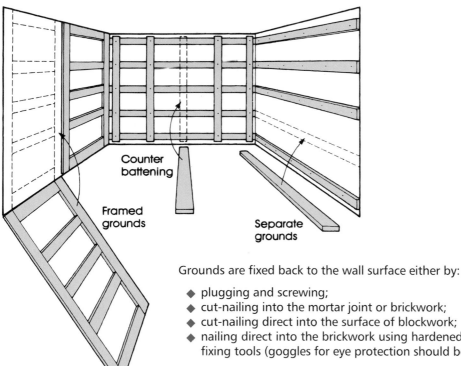

Counter
battening

Framed
grounds

Separate
grounds

Grounds are fixed back to the wall surface either by:

◆ plugging and screwing;
◆ cut-nailing into the mortar joint or brickwork;
◆ cut-nailing direct into the surface of blockwork;
◆ nailing direct into the brickwork using hardened nails or by using cartridge-fixing tools (goggles for eye protection should be worn).

They must be plumbed and lined in to provide a flat surface (see Figure 4.116).

Figure 4.115 *Types of grounds*

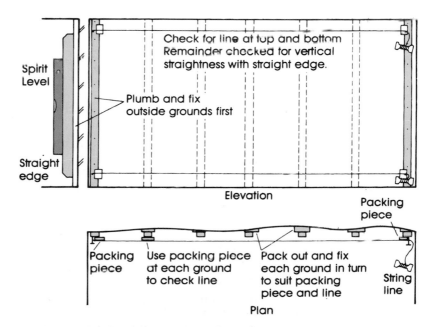

Spirit
Level

Check for line at top and bottom
Remainder checked for vertical
straightness with straight edge.

Plumb and fix
outside grounds first

Straight
edge

Elevation

Packing
piece

Packing
piece

Use packing piece
at each ground
to check line

Pack out and fix
each ground in turn
to suit packing
piece and line

String
line

Plan

Figure 4.116 *Plumbing and lining grounds*

Fixing to grounds – the fixing of panelling to grounds should as far as possible be concealed (see Figure 4.117).

Panelling should be lowered into position and held in place by interlocking grounds, one fixed to the wall and the other to the panelling.

Interlocking metal plates or keyhole slots and screws also provide a fixing when the panelling is lowered into position. Cover fillets or other trim may be used to conceal panelling that has been surface screwed.

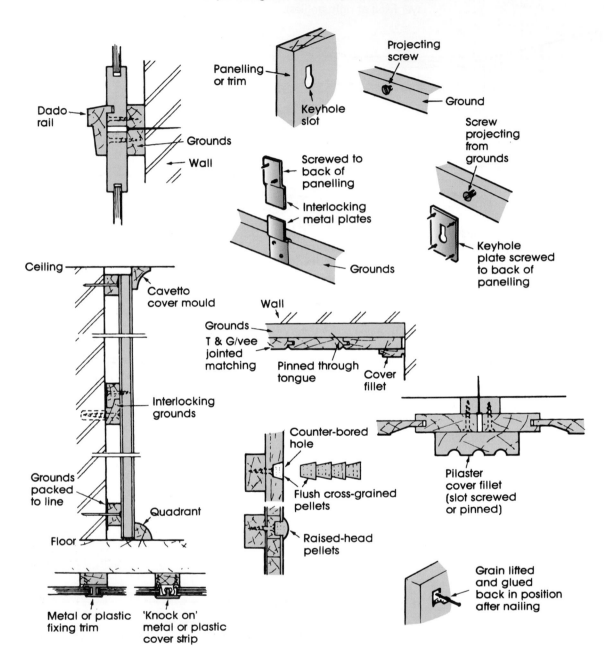

Figure 4.117 *Concealed fixings*

Corner details – the method of forming internal and external angles will depend on the type of panelling, but in any case they should be adequately supported by grounds fixed behind. Figure 4.118 shows various details. Tongued-and-grooved joints, loose tongues, rebates or cover fillets and trims have been used to locate the panelling members and at the same time conceal the effects of moisture movement.

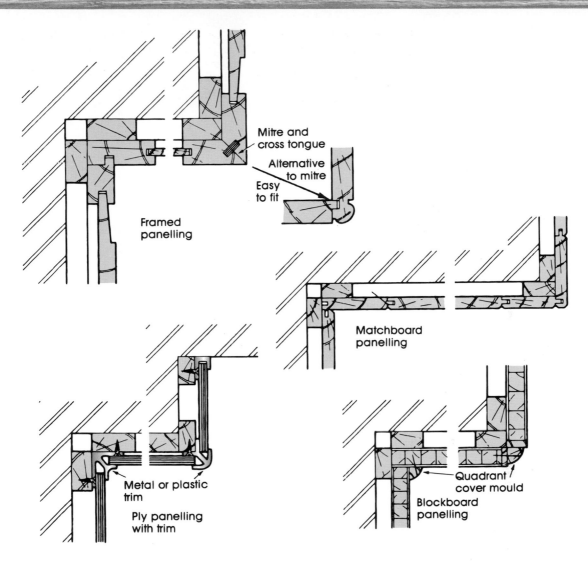

Mitre and
cross tongue

Alternative
to mitre

Easy
to fit

Framed
panelling

Matchboard
panelling

Metal or plastic
trim

Ply panelling
with trim

Quadrant
cover mould

Blockboard
panelling

Figure 4.118 *Corner details*

Where matchboarding or similar timber strips are used for paneling, they are fixed back to grounds using secret nails through the tongue. The boards at either end of a wall should be of equal width and may be surface nailed (see Figure 4.119).

A simple calculation can be carried out to determine the required width.

example

A 3.114m length of wall is to be panelled with 95mm (90mm covering width) matchboard.

Divide the length of wall in millimetres by the covering width of the board.

3114 ÷ 90 = 34.6 boards

Therefore 35 boards are required = 33 whole boards and two end boards.

Width of cut end boards = 1.6 × 90 ÷ 2

= 72mm

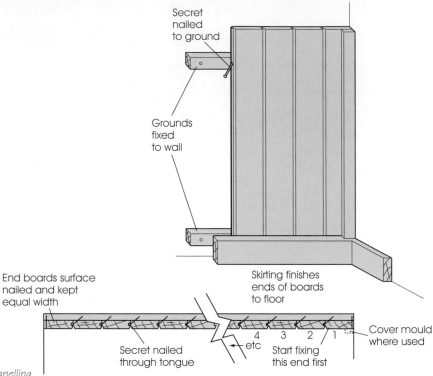

Secret nailed to ground

Grounds fixed to wall

End boards surface nailed and kept equal width

Skirting finishes ends of boards to floor

Secret nailed through tongue

4 3 2 1
etc   Start fixing this end first

Cover mould where used

Figure 4.119 *Layout of matchboard panelling*

General requirements of panelling

1. Before any panelling commences it is essential that the wall construction has dried sufficiently.
2. All timber should be of the moisture content required for the respective situation (equilibrium moisture content, M/C).
3. The backs of the panelling sections should be sealed prior to fixing, thus preventing moisture absorption.
4. Timber for grounds should be preservative treated.
5. A ventilated air space is desirable between the panelling and the wall.
6. Provision must be made for a slight amount of moisture movement in both the panelling sections and the trim.
7. The positioning of the grounds must be planned to suit the panelling.
8. The fixing of the panelling to the grounds should be so designed that it is concealed as far as possible.

Cladding

Cladding is the non-load-bearing skin or covering of external walls for weathering purposes, e.g. timber boarding, sheet material, tile hanging and cement rendering.

Timber cladding for either timber-framed buildings or those of brick or blockwork construction is fixed to battens or grounds spaced at a maximum of 600 mm centres. A moisture barrier is fixed below the cladding to battens in timber-framed buildings to provide a second line of protection to any wind-driven rain that might penetrate the cladding. This is often termed a 'breather paper', as it must allow the warm air vapour to pass or breathe through it from inside the building and not get trapped in the wall. The moisture barrier is often omitted for claddings over brick or blockwork.

The battens are fixed to the studs of the timber frame or direct to the brick or block surface. They must be lined and levelled to provide a flat surface.

Cladding is normally specified as 16mm. Feather-edged boards will taper to about 6mm at their thin edge. Natural durable timber cladding such as Western Red Cedar may be used without preservative treatment or any subsequent finish. Most other softwood claddings are not naturally durable and must be preservative treated.

It is recommended that all timber used for cladding, the grounds as well as the boarding, is preservative treated before use. There is little point in treating the face of cladding after it has been fixed, leaving the joints or overlapping areas, back faces and grounds untreated. Any preservative treated timber cut to size on site will require re-treatment on the freshly cut ends and edges. This can be carried out by applying two brush flood coats of preservative.

Venting the space behind the cladding is important in order to reduce the risk of timber decay. Venting allows any moisture vapour passing through the wall from the inside, or any moisture absorbed by the cladding from the air to be safely dispersed, see Figure 4.120. Venting of the cavity can be considered as an increased fire hazard and cavity barriers may require fitting in order to prevent the passage of smoke and flames. Again, approval from the Building Inspector will be required.

did you know?

You should only use one nail per board width in order to reduce the likelihood of splitting after moisture movement.

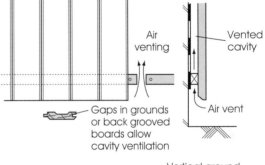

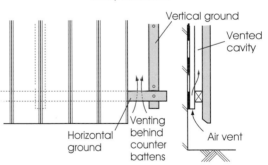

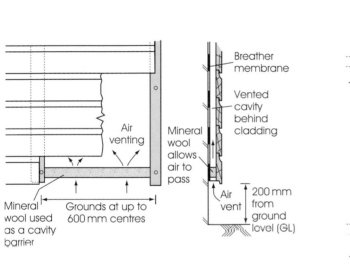

Figure 4.120 *Venting the cavity*

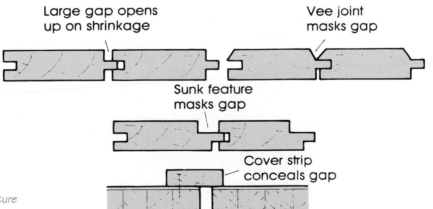

Figure 4.121 *Effects of moisture movement and the remedy*

Timber is a hygroscopic material, meaning it readily absorbs or gives off moisture to achieve a balance with its surroundings. In doing so it expands or shrinks. Cladding sections should be designed to minimize, mask or cover any unsightly gaps resulting from moisture movement (see Figure 4.120).

Plywood used for cladding must be WBP rated. This means that its veneer layers have been glued using a weather and boil proof adhesive to give it a very high resistance to all weather conditions.

Plastic (PVCu) cladding sections are also used, having the advantage over timber of requiring less maintenance. Cutting and fixing these profiles is the same as for timber.

Cladding details

Careful detailing is required where claddings finish around door and window openings or where different claddings join: see Figure 4.122. This is to allow for differing movement and prevent weather/insect penetration.

Cladding fixings

Cladding fixings are normally nails at least 2.5 times in length of the cladding's thickness. Ferrous metal (metal that will rust) nails should be galvanized or sherardized to resist corrosion. Copper or aluminium nails must be used with Western Red Cedar as it accelerates rusting in ferrous metal and causes unsightly timber staining. Stainless steel nails are recommended to be used with PVCu profiles again to prevent staining.

Boards should be single nailed to each batten and care must be taken to ensure the nails do not go through the board below where they overlap as splitting will occur when the cladding is subjected to moisture movement (see Figure 4.123).

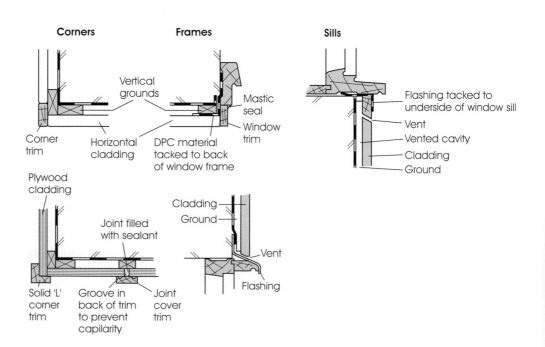

Figure 4.122 *Typical cladding details*

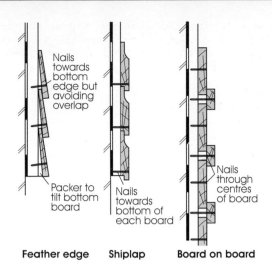

Feather edge   **Shiplap**   **Board on board**

Nails towards bottom edge but avoiding overlap

Packer to tilt bottom board

Nails towards bottom of each board

Nails through centres of board

Figure 4.123 *Typical cladding fixings*

measuring up

30. Produce a sketch to show the difference between dado and frieze height panelling.

31. Explain why grounds must be plumbed and lined in.

32. Explain why exterior plywood cladding should be specified as WBP rated.

33. Explain the reason for using copper nails when fixing Western Red Cedar cladding.

34. State the purpose of the vee joint in matchboard panelling.

35. State the purpose of incorporating a moisture barrier below external timber cladding.

36. State a reason why Western Red Cedar may be specified for external cladding.

37. Produce a sketch to show how wall panelling may be secret fixed using interlocking grounds.

38. Name THREE means of secret fixing panelling or matchboarding.

39. State the reason for venting the cavity behind cladding.

Second Fixing

Chapter 4

Maintenance

This chapter is intended to provide the reader with an overview of maintenance work. Its contents are assessed in the NVQ Unit No. VR 12 Maintain Non-structural Carpentry Work.

In this chapter you will cover the following range of topics:

◆ defining maintenance;
◆ agents of deterioration;
◆ undertaking repairs and maintenance work;
◆ timber repairs;
◆ brickwork and plastering repairs;
◆ replacing damaged tiles;
◆ glazing and painting;
◆ guttering and downpipes.

What's required in VR 12 Maintain Non-structural Carpentry Work?

To successfully complete this unit you will be required to demonstrate your skill and knowledge of the following maintenance activities:

◆ repair or replacement of non-structural carpentry work;
◆ making good defective brickwork and plastering;
◆ glazing and painting;
◆ replacement of gutters and downpipes.

You will be required practically to:

◆ repair and/or replace frames;
◆ repair and/or replace mouldings;
◆ repair and/or replace door and/or window ironmongery;
◆ repair and/or replace guttering and downpipes;
◆ repair and/or replace sash cords;
◆ prime repairs made to woodwork;
◆ make good defective plasterwork and/or brickwork, caused by the undertaking of repairs.

Note: Except for sash cords, plasterwork and brickwork, for the successful completion of all other practical tasks evidence must be work based.

Defining maintenance

It is an accepted fact that all buildings will deteriorate (develop faults and defects, which if not rectified may lead on to failures) to some extent as they age. This deterioration may even start as the individual components are incorporated into the building elements during the construction process. In certain circumstances the deterioration of the components may have started either prior to their delivery to the building site or during the storage, before the commencement of construction operations.

The rate and extent to which a building deteriorates is dependent on one or more of the following main factors: maintenance; the environment; design and construction.

Maintenance is taken to mean the keeping, holding, sustaining or preserving of a building and its services to an acceptable standard. This may take one of two forms: planned maintenance or unplanned maintenance.

Planned or routine maintenance
This is a definite programme of work aimed at reducing to a minimum the need for often costly unplanned work. It includes:

◆ the annual inspection and servicing of general plumbing, heating equipment, electrical and other services etc.;
◆ the periodic inspection and cleaning out of gutters, gullies, rainwater pipes and airbricks etc.;
◆ the periodic redecoration, both internally and externally;
◆ the routine general inspection/observation of the building fabric and moving parts.

Preventative maintenance
Also included under this heading is what is known as preventative maintenance. Basically this is any work carried out as a result of any of the previous inspections in anticipation of a failure, e.g. the early replacement of an item, on the assumption that minor faults almost certainly lead onto bigger and more costly faults unless preventative work is carried out.

Unplanned emergency or corrective maintenance
This is work that is left until the efficiency of the element or service deteriorates below the acceptable level or even fails altogether. This is the most expensive form of maintenance, making inefficient use of both labour and materials and often also creating serious health/safety risks, and is the type most often carried out. This is because the allocation of money to enable maintenance work to be planned is often given low priority.

Environmental factors
These include:

◆ the deterioration of components and finishes owing to chemical pollution in the atmosphere;
◆ the effect of the elements (weather) on the structure, e.g. frost, rain, snow, sun and wind;
◆ the effect of these elements when allowed to penetrate into the building;
◆ the deterioration of components owing to biological attack (fungal decay and insect attack).

Design and construction factors
Faulty design and construction methods can lead to rapid deterioration of a building. In fact over 30% of all maintenance/repair could be avoided if sufficient care is taken at the design and construction stages.

Faulty design
This results from inadequate knowledge or attention to detail on the part of the architect or designer leading to, for example, poor specification of materials/components, structural movement, moisture penetration, biological attack and the inefficient operation of building services.

Faulty construction

Inadequate supervision during the construction process can result in poor workmanship, the use of inferior materials and the lack of attention to details/ specifications. These can all lead to the same problems as those stated for faulty design, resulting in subsequent problems and expense for the building owner.

Agents of deterioration

Apart from the natural ageing process of all buildings during their anticipated life (however well maintained), deterioration of buildings can be attributed directly to one or often a combination of the following agents:

◆ dampness;
◆ movement;
◆ chemical attack;
◆ biological attack;
◆ infestation.

Dampness

Dampness in buildings is the biggest single source of trouble. It causes the rapid deterioration of most building materials, can assist chemical attack and creates conditions that are favourable for biological attack. Dampness can arise from three main external sources: rain penetration, rising damp and condensation. Leaking plumbing and heating systems and spillage of water in use are also significant causes of dampness.

Rain penetration

This is rain penetrating the external structure either through the walls or the roof and appearing on the inside of the building as damp patches. After periods of heavy rain these patches will tend to spread and then dry out during prolonged periods of dry weather. They will, however, never completely disappear, as a moisture stain and in some cases even efflorescence (crystallized mineral salts) will be left on the surface.

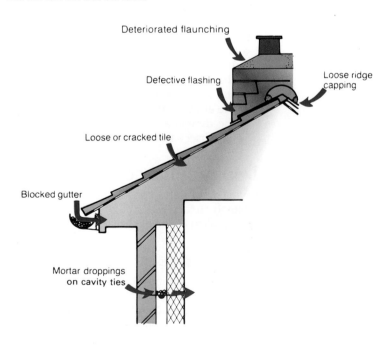

Figure 5.1 *Rain penetration*

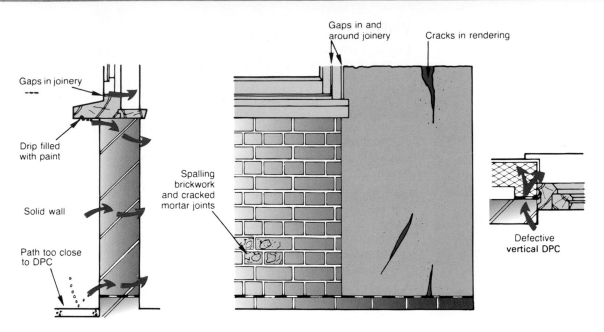

Figure 5.2 *Rain penetration*

Mould growth (fungi resulting in dark-green or black patchy spots) may occur in damp areas particularly behind furniture, in corners and other poorly ventilated locations. The main causes of rain penetration are shown in Figures 5.1 and 5.2. It can be seen that penetration takes places through gaps, cracks, holes and joints in, around or between components and elements.

Roofs

Loose or missing tiles or slates including the hip and ridge capping tiles will allow rainwater to run down rafters, causing damp patches on the ceilings and tops of walls. These patches may appear some distance away from the defective area as the water spreads along timbers and across the ceiling etc. This dampness will also saturate any thermal insulation material, making it ineffective. If left unrepaired, saturation of the roof timbers will occur, leading to fungal decay in due course. Another major area of penetration is around the chimney stack and other roof-to-wall junctions; this may be due to cracked chimney pots, cracked or deteriorating flaunchings (the sloping mortar into which the pots are set), or corroded or pitted metal flashings (these cover the joint between the stack or wall and the roof), which may be cracked or deteriorated. Poor pointing to the stack can also be a cause of penetration. Any of these defects can cause large patches of damp on the internal wall.

Maintenance **Chapter 5**

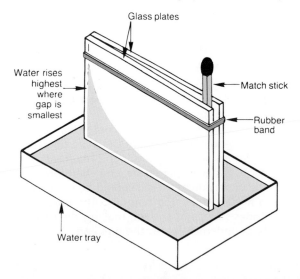

Figure 5.3 *Experiment to show capillarity*

Chapter 5   Maintenance

Walls

Clearly rainwater travels downwards and when assisted by high winds it will travel sideways through gaps. But depending on the nature of the material, it can often move unassisted sideways or upwards because of capillary attraction.

There are two main conditions that promote capillarity in the external envelope. The fine cellular structure of some materials provides the interconnecting pores through which water can travel, also the fine joints between components, e.g. wall and door or window frame, mortar joints between brickwork, close joints between overlapping components. The risk of capillarity is reduced or avoided by either:

◆ physically separating the inside and outside surface by introducing a gap (e.g. cavity wall construction);
◆ introducing an impervious (waterproof) barrier between components (e.g. mastic pointing, damp-proof courses (DPCs) in walls, damp-proof membranes (DPMs) in floors, moisture barriers behind cladding and flashing at wall-to-roof intersections).

Over time, the water resistance of brick/stonework and their mortar joints will deteriorate. This deterioration can be accelerated by the action of frost. Rainwater may accumulate below the surface and freeze. Ice expands causing the brickwork/stonework and its mortar joints to spall (crumble away). The wall then offers little resistance to the weather and should be replaced. This entails either:

◆ cutting the surface of the spalled components back and replacing with matching thin components (half bricks) and finally repointing the whole wall;
◆ hacking off (cutting back to remove spalling) the entire wall and covering with one of the standard wall finishes, cement rendering, rough cast, pebble dash, Tyrolean or silicone-nylon fibre.

Cracks in cement rendering and other wall finishes can be caused by shrinkage on drying, building movement or chemical attack. Once opened up, deterioration is accelerated by frost action. Small cracks may be enlarged and filled with cement slurry. Large areas that may have 'blown' (come away) from the surface will require hacking back to sound (firmly adhering) work and replaced. With cavity wall construction, rainwater that does penetrate the outer leaf should simply run down inside the cavity and not reach the internal leaf. The vertical mortar joints of the outer leaf are sometimes raked out at intervals along the bottom of the cavity to provide weep holes through which the water can escape.

However, when the cavity is bridged by a porous material (e.g. the collection of mortar droppings on the wall ties during construction), the water will reach the inner leaf causing small, isolated damp patches on the internal wall surface. The remedy for this fault is to remove one or two bricks of the outer leaf near the suspected bridge and either clean out or replace the tie as necessary.

Dampness around door and window frames is likely to be caused by wind-assisted rain entering the joint between the wall and frame, by the action of capillarity or by a defective vertical DPC used around openings in cavity walls, where the inner and outer leaf join. An exterior mastic can be used to seal the joints, but where DPCs are defective they will require cutting out and replacing. A check should be made at the sill level of frames. Cracked sills allow water to penetrate and therefore should be filled. The drip groove on the underside of the sill should be cleaned out as it often collects dirt/dust and is filled by repeated painting. The purpose of the drip groove is to break the under surface of the sill, making the water drip off at this point and not run back underneath into the building.

did you know?

Capillary attraction is the phenomenon whereby a liquid can travel against the force of gravity, even vertically, in fine spaces or between surfaces close together, due to its own surface tension. The smaller the space, the greater the attraction. A simple experiment that demonstrates capillary attraction is illustrated in Figure 5.3. It shows that the water rises highest where the two glass plates touch and reduces as the gap widens. Measures taken to prevent or reduce the effect of capillary attraction are called 'anti-capillary' measures. The grooves that run around the edges of window frames to widen the gap are an example of an aniti-capillary measure.

Blocked or cracked gutters and downpipes, dripping outside taps and constantly running overflows can cause an excessive concentration of water in one place, which will be almost permanently damp. This will result in an accelerated deterioration of the wall and subsequent internal damp patches etc. The immediate fault can be easily rectified by repairing or replacing the defective component. But if left unattended the resulting damage to the building structure has most serious and costly implications.

Rising damp

This is normally moisture from below ground level rising and spreading up walls and through floors by capillarity. This most often occurs in older buildings. Many of these were built without DPCs and DPMs or, where they were incorporated, have broken down possibly with age (e.g. slate, a one time popular DPC material cracks with building movement, thus allowing capillarity). The visual result on the walls is a band of dampness and staining spreading up from the skirting level; wallpaper peeling from the surface and signs of efflorescence. The skirting, joists and floorboards adjacent to the missing or failed DPC are almost certain to be subject to fungal attack. Solid floors may be almost permanently damp causing considerable damage to floor coverings and adjacent timber/furniture etc. Rising damp can still occur in buildings that have been equipped with DPCs and DPMs (see Figure 5.4).

did you know?

Damp-proof courses are normally located at least two courses of brickwork (150 mm) above the adjacent ground level. This is because even very heavy rain is unlikely to bounce up and splash the walls much more than 100 mm from the surrounding surface. Thus the splashed rainwater is still prevented from rising above the DPC. Where the surrounding surface is later raised these splashes might bypass the DPC and result in rising damp.

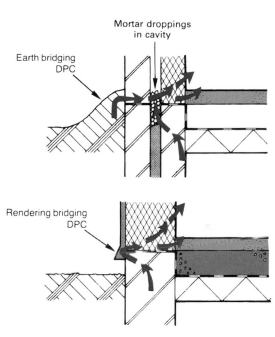

Figure 5.4 *Rising damp in walls*

One of the main reasons for this is the bridging of DPCs; in the case of cavity walls, builders' mortar droppings or rubble may have collected at the bottom of the cavity, allowing moisture to rise above the DPC level; or earth in a flower bed being too high above the DPC.

Weak porous rendering that has been continued over the DPC is another means by which the DPC may be bypassed.

Maintenance

Chapter 5

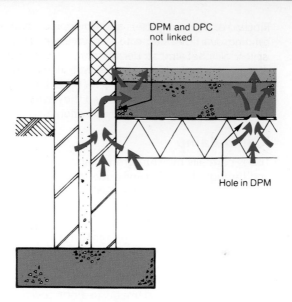

DPM and DPC
not linked

Hole in DPM

Figure 5.5 *Rising damp in solid ground floors*

In solid floors with a DPM, rising damp can only occur if this is defective (see Figure 5.5). For example, it may have been penetrated by jagged hardcore during the pouring of the over-site concrete or have been inadequately lapped (permitting capillarity between the lapped joint), or finally it may not have been linked in with the DPC in the surrounding walls (allowing moisture to bypass at this point).

The remedy to rising damp faults will, of course, vary; bridged or bypassed DPCs can be rectified by simply removing the cause, e.g. lowering the ground level or removing mortar and rubble from the cavity etc. Where the DPC itself is faulty or missing altogether, one can be inserted by either cutting out a few bricks at a time to allow the positioning of a new DPC, sawing away the mortar joint a section at a time and inserting a new one.

Alternatively, liquid silicone may be injected near the bottom of the wall. This soaks into the lower courses, which then act as a moisture barrier preventing capillarity.

Localized faults in DPMs can be remedied by cutting out a section of the floor larger than the damp patch, down to the DPM, taking care not to cut through it. This should reveal the holed or badly lapped portion, which can be repaired with a self-adhesive DPM. An alternative method that can also be used in floors without any DPM is to cover the existing concrete floor with a liquid bituminous membrane or a sheet of heavy-duty polythene sheeting before laying a new floor finish, although to be effective it should be joined into the DPC.

Condensation

The results of this form of dampness are often mistakenly attributed to rain penetration or rising damp, as they can all cause damp patches, staining, mould growth, peeling wallpaper, efflorescence, the fungal attack of timber and generally damp, unhealthy living conditions. The water or moisture for condensation actually comes from within the building: people breathing, kettles boiling, food cooking, clothes washing and drying, bath water running, etc. Each of these processes adds more moisture to the air in the form of vapour.

Air is always capable of holding a certain amount of water vapour. The warmer the air, the more vapour it can hold, but when air cools the excess vapour will revert to water. This process is known as condensation. Thus whenever warm moist air meets a cool surface condensation will occur (see Figure 5.6). This can only be controlled effectively by achieving a proper balance between heating, ventilation and insulation. The building should be kept well heated but windows should be opened or mechanical ventilators used especially in kitchens and bathrooms, to allow the vapour-laden air to escape outside and not spread through the building. External walls need thermally insulating to remove their cold surfaces. Both cavity wall insulation and lining the walls with a thin polystyrene veneer help a great deal. Double-glazed windows also help reduce condensation by preventing the warm moist air coming into direct contact with the cold outside pane of glass.

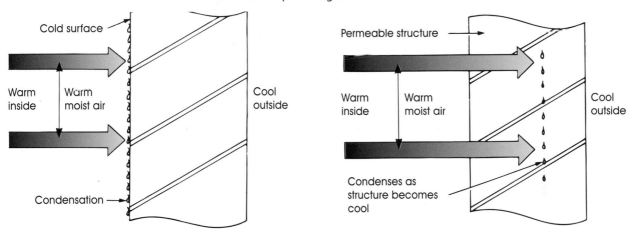

Figure 5.6 *Surface condensation*

Figure 5.7 *Interstitial condensation*

In addition to this surface condensation, there is another condensation problem that occurs when wall surfaces are warm. This is known as interstitial or internal condensation. This is illustrated in Figure 5.7. It is caused by the warm, moist air passing into the permeable structure until it cools, at which point it condenses, thus leading to the same problems associated with penetrated and rising dampness.

Interstitial condensation can be dealt with either:

◆ by the use of a vapour barrier (this prevents the passage of water vapour) on the warm inside of the wall, e.g. a polythene sheet or foil backed plasterboard; or
◆ by allowing this water vapour to pass through the structure into a cavity where it can be dispersed by ventilation.

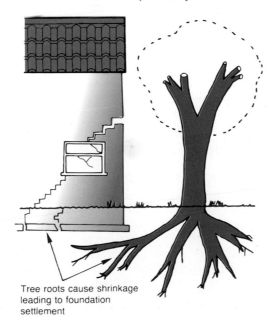

Tree roots cause shrinkage leading to foundation settlement

Figure 5.8 *Ground movement*

Movement

The visual effects of movement (Figure 5.8) in buildings may apparently be of a minor nature, e.g. windows and doors that jamb or bind in their frames; fine cracks externally along mortar joints and rendering; fine cracks internally in plastered walls and along the ceiling line etc. They can however be the first signs of serious structural weakness. Movement in buildings takes two main forms, these being: ground movement and movement of materials.

Ground movement

Any movement in the ground will cause settlement in the building. When it is slight and spread evenly over the building it may be acceptable, although when it is more than slight or is differential (more in one area than another), it can have serious consequences for the building's foundations and load-bearing members, requiring expensive temporary support (shoring) and, subsequently, permanent underpinning (new foundations constructed under existing ones).

Ground movement is caused mainly by its expansion and shrinkage near the surface, owing to wet and dry conditions. Compact granular ground suffers little movement, whereas clay ground is at high risk. Tree roots cause ground shrinkage owing to the considerable amounts of water they extract from it. Tree roots can extend out in all directions from the tree's base, greater than its height. Overloading a structure beyond its original design load can also result in ground movement.

Frost also causes ground movement. Water in the ground expands on freezing. Where it is allowed to expand on the undersides of foundations it has a tendency to lift the building (known as frost heave) and drop it again on thawing. This repeated action often results in serious cracking. Freezing of ground water is limited in the UK to about the top 600 mm in depth.

Movement in materials

All building materials will move to some extent owing to one or more of the following reasons: temperature changes, moisture-content changes and chemical changes. Provided the building is designed and constructed to accommodate these movements or steps are taken to prevent them, they should not lead to serious defects.

Temperature changes – these cause expansion on heating and shrinkage on cooling; metals and plastics are particularly affected, although concrete, stonework, brickwork and timber can also be affected.

Moisture changes – many materials expand when wetted and shrink on drying. This is known as moisture movement. The greatest amount of moisture movement takes place in timber, which should be painted or treated to seal its surface. Brickwork, cement rendering and concrete can also be affected by moisture movement. Rapid drying of wetted brickwork in the hot sun can result in cracks, particularly around window and door openings.

Chemical and fungal attack

Corrosion and sulphate attack

Corrosion causes metals to expand and lose strength. Corrosion of steel beams can lift brickwork causing cracks in the mortar joint. Bulges in cavity brickwork may be caused by corroded wall ties. The sulphate attack of cement is either in the ground or from products of combustion in chimneys. The sulphate mixes with water and causes cement to expand. Sulphate-resisting Portland Cement (SRPC) should be used in conditions where high levels of sulphate are expected.

Smoke containing chemicals is given off into the atmosphere as a result of many manufacturing processes. This mixes with water vapour and rainwater to form dilute or weak acid solutions. These solutions corrode iron and steel, break down paint films and erode the surfaces of brickwork, stonework and tiles. The useful life of materials in these environments can be prolonged by regular cleaning to remove the contamination.

Ageing – exposure to sunlight can cause bleaching, colour fading of materials and even decomposition owing to solar radiation. Particularly affected are bituminous products, plastics and painted surfaces.

Biological attack

Timber, including structural, non-structural and timber-based manufactured items are the targets for biological attack. The agents of this are fungi and wood-boring insects. Given the right conditions an attack by one or both agents is almost inevitable.

There are two main types of fungi that cause decay in building timbers, these being dry rot and wet rot.

Dry rot

Dry rot is more serious and is more difficult to eradicate than wet rot. It is caused by a fungus that feeds on the cellulose found mainly in sapwood (outer layers of a growing tree). This causes timber to lose strength and weight, develop cracks in brick-shaped patterns and finally to become so dry and powdery that it can easily be crumbled in the hand. The appearance of a piece of timber after an attack of dry rot is shown in Figure 5.9. Two initial factors for an attack are damp timber in excess of about 20% moisture content (MC) and bad or non-existent ventilation.

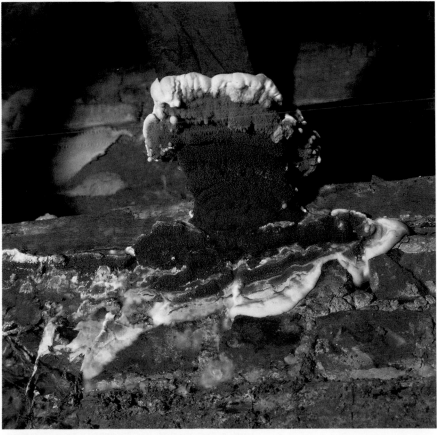

Figure 5.9 *Timber after dry rot attack*

did you know?

The cracking of timber into brick-shaped patterns as a result of dry rot is termed 'cuboidal cracking'.

As the fungus is a living plant, an attack commences with the germination of its microscopic spores (seeds) that send out hyphae (roots) into the timber to feed on the cellulose. Once established, these hyphae branch out and spread through and over the timber forming a matt of cottonwool-like threads called mycelia. At this stage, the hyphae can penetrate plaster and brickwork in search of further timber supplies to feed on. This further timber supply need not be damp as the developed hyphae can conduct their own water supply, thus adjusting the moisture content as required. Finally the fruiting body, like a fleshy pancake with an orange-brown centre, will start to ripen and eject into the air millions of the rust-red spores, to begin the process elsewhere. Very often in the early stages, apart from a damp, musty, mushroomy smell, there is little evidence of an attack. It is not until the wall panelling, skirting or floorboards are removed that the full effects are realized. An advanced state of dry rot is skirting and flooring is shown in Figure 5.10.

Figure 5.10 *Advanced dry rot*

Eradication and treatment – to eradicate an attack of dry rot, first rectify sources of dampness and bad ventilation:

◆ Remove all traces of decayed timber and at least 600 mm of apparently sound timber beyond the last signs of attack.
◆ All affected timber including swept-up dust, dirt and old wood shavings etc. must be sealed in airtight polythene bags and arrangements made for their incineration (contact your local authority for information). This prevents spreading and kills hyphae and spores.
◆ Strip plaster from walls, wire brush brickwork, heat up brickwork with a blow lamp to sterilize and brush or spray wall with a dry-rot fungicide (this kills any hyphae and spores in the walls).
◆ Finally, work may be reinstated with preservative treated timber.

Wet rot

Wet rot is also caused by a fungus, but it does not normally involve such drastic eradication treatment, as it does not spread to the same extent as dry rot. It feeds on wet timber (30% to 50% MC) and is most often found in cellars, neglected external joinery, ends of rafters, under leaking sinks or baths and under impervious (waterproof) floor coverings. During an attack, the timber

did you know?

The idea behind preservative treatment is to poison the food supply of fungi and wood-boring insects, by applying a toxic liquid to the timber.

becomes soft, darkens to a blackish colour and develops cracks along the grain. Very often timber decays internally with a fairly thin skin of apparently sound timber remaining on the surface. The hyphae when apparent are dark brown or black; internally hyphae may be white and form into sheets. Its fruiting body, which is rarely found, is of an irregular shape and normally olive green in colour, as are the spores. Figure 5.11 shows the appearance after an attack of wet rot in the rafters of a roof.

safety tip

Remember that when using preservatives or other chemicals you must always follow the manufacturer's safety instructions concerning their use and the appropriate PPE to be worn.

Figure 5.11 *Wet rot in rafters*

Timber treatment – to eradicate an attack of wet rot all that is normally required is to cure the source of wetness and allow the timber to dry out. Replacement of soft timber may be required after an extensive attack, particularly where structural timber is concerned. Non-structural timber may be treated with a wet rot wood hardening fluid.

Maintenance **Chapter 5**

Name	Actual size	Location and timber attacked
Furniture beetle		Softwoods and the sapwood of hardwoods; causes considerable damage to timber, flooring and furniture
Death-watch beetle		Mainly hardwoods in old damp buildings (churches); often in association with fungal attack
Lyctus beetle (powder post)		Sapwood of freshly-cut hardwoods; normally in timber yards before use
House long-horn beetle		Sapwood of softwoods; mainly roof timbers
Weevils		Damp or decayed hardwoods and softwoods; often found around sinks, baths, toilets and in cellars

Table 5.1 *Wood-boring insects*

Infestation

Wood-boring insects

This is also known as woodworm, after the larvae that are able to feed on, and digest, the substance of wood. The majority of the damage done to building timber in the UK can be attributed to five species illustrated in Table 6.1, which also includes their identifying characteristics.

The female adult beetle lays eggs during the summer months, usually in the end grain, open joints or cracks in the timber. This affords the eggs a certain amount of protection until the larvae hatch. The larvae then start their damaging journey by boring into the timber, consuming it and then excreting it as fine dust. The duration of this stage varies between six months and 10 years, depending on the species. During the early spring, at the close of this stage, the larvae bore out a pupal chamber near the timber surface, where they undergo the transformation into adult beetles. This takes a short period after which the beetles bite out of the timber leaving characteristic flight holes. The presence of flight holes is often the first external sign of an attack. After emerging from the timber the beetle's instinct is to mate, lay eggs and then die, thus completing one life cycle and starting another.

Timber treatment – to eradicate an attack of wood-boring insects open up the affected area (take up floorboards etc.), remove all affected timber and replace with new preservative-treated timber.

All removed timber and swept-up dust and old wood shavings etc. must be sealed in airtight polythene bags and arrangements made for their incineration.

Brush timber to remove dust, strip off surface coating, e.g. paint, varnish etc. (woodworm fluid will not penetrate surface coatings). Apply two coats of a proprietary woodworm killer by brush or spray to all timber, even apparently unaffected timber. Pay particular attention to cracks, joints, end grain and flight holes. Inspections for fresh flight holes should be carried out for several successive summers. A further treatment of fluid will be required if any are found. (Fresh bore dust around the affected area indicates fresh flight holes.)

measuring up

1. Define the term 'building maintenance'.

2. Name two factors that affect the rate and extent to which a building deteriorates.

3. Name three agents to which deterioration in buildings can be directly attributed.

4. Produce a sketch to show two methods by which moisture may bypass DPCs.

5. Define the term 'capillary attraction'.

6. Name two causes of movement in buildings and identify their likely effect.

7. State the purpose of treating timber with preservative.

8. Identify two causes of rising damp and suggest a remedy for each.

9. List two defects under each of the following headings that can lead to the rapid deterioration of a building:
    (a) movement
    (b) biological attack.

10. Identify the probable causes of the following defects:
    (a) small isolated damp patches at intervals on the internal leaf of an external cavity wall
    (b) small holes in the surface of timber with fine dust around them
    (c) damp patch in the centre of a solid ground floor.

Maintenance Chapter 5

Undertaking repairs and maintenance work

Building firms that specialize in maintenance work will often tend to employ or give preference to operatives, who possess multi-skills, as they will be expected to carry out, in conjunction with their main craft skill, a range of basic skills of the other crafts, for example:

◆ After hanging a replacement door you may be expected to paint it
◆ When replacing a window you may be expected to glaze it, re-lay the brickwork under the sill and patch in the plasterwork. If this was in a bathroom or kitchen it may also involve replacing ceramic tiles.

Inspections and repair surveys

Whenever a building firm undertakes major repairs or maintenance it is desirable to undertake a survey of the building. The extent of the measurements, sketches and details taken will depend on the nature and extent of the work. Clearly a survey prior to replacing windows in a house will be very different from that of one involving structural movement.

Existing information

In many cases, there will already be in existence information that can assist you when carrying out a survey:

◆ **_Drawings_** – make enquiries of the building's owners to determine whether there are any existing drawings of the building. These may be in their possession from when the building was new or from when an extension was added some time in the past. Figure 5.12 illustrates the typical general location plans from when the house was built, which might be available. If so these can simplify your task by forming the basis of the survey sketches.

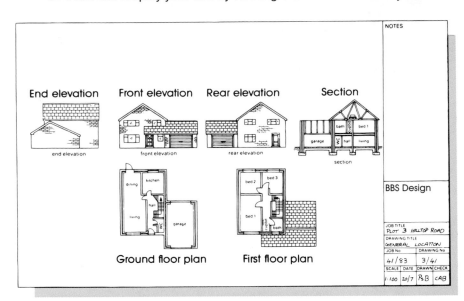

Figure 5.12 _General location plans_

◆ **_Previous survey reports_** – often there are existing reports, made by building society surveyors for mortgage purposes, or structural reports made for insurance purposes when a claim is being made. Figure 5.14 shows extracts from a typical schedule of remedial works made following a structural survey of a house, as part of an insurance claim. This identifies the main areas of concern. You would concentrate on these points to estimate the precise amount of work involved.

Undertaking the survey

Before you start the actual survey you should provisionally look the building over, both internally and externally, to determine its general layout and any likely difficulties. This will involve making notes and sketches to create a true record of the building's existing condition, including work/defects outside your craft.

Internal survey

Sketch plans are made of each floor or room, starting at ground floor level. You may be guided to the apparent major problem by the building's occupants, or another survey report etc. (e.g. schedule of remedial work). This can be the

SCHEDULE OF REMEDIAL WORKS
TO THE SUPERSTRUCTURE

Mrs J Hill
26 Thorneywood Road
Long Eaton
Nottingham

A.   *Brief Description of the Works*

*In describing the property all
left and right assume that the
Thorneywood Road.*

*The property has suffered fro
movement. The foundations c
purposes stable but slight arf
due to normal structural mov
the established crack lines.
therefore designed to minimi
disguise that which is unavoid*

B.   *General Specification (to app
Indicated otherwise in the Sc*

1.   GENERAL
Unless specifically indic
and workmanship in the re
to those in the existing pr
sufficient to comply with t
whichever is higher, or as
of this schedule.

2.   REPAIR OF CRACKS IN EXTE
Rake out cracks in mor
of 50mm, drive in slate we
mortar and repoint with c
colour and texture to mat
brickwork to be re-decora
bricks are to be removed
bonded with resin adhesiv

3.   REPAIR OF CRACKS IN EXTERNAL RENDERING
Remove loose rendering from around the cracked
area and bonded rendering within 150mm on each side
of the cracks, repair the substrate as external brickwork
and reinstate the rendering incorporating light
reinforcement and with a texture to match the existing.
Repairs in painted rendering are to be redecorated with a
minimum of 2 coats of resin-based masonry paint to
match the existing with the redecoration extending over
the complete panel containing the repair or to such other
limits as necessary to avoid a mismatch with existing.

4.   REPAIR OF CRACKS IN INTERNAL WALLS
Remove all loose pl
bonded plaster to a d
side of the crack.
external brickwork and
scrim reinforcement. C
removed and replaced

5.   REPAIR OF CEILING CRAC
(a)   Minor cracks up t
Rake out ligh
decoration
(b)   Cracks 0.5mm –
Rake out, ta
smooth surface t
(c)   Cracks larger than
Underboard, to
the complete ceili

6.   EASING DOORS AND W
Prices against item
redecoration should a
and windows as neces

7.   REDECORATION
Use movement-tolerant material (Superglypta or
similar) unless specified otherwise. If materials are
used which are not in accordance with the
specification the property owner must be made
aware of the increased risk of recurrent damage.
Walls and ceilings containing repaired cracks to be
lined with linen-backed paper. Artex or similar finishes
are to be reinstated as existing. Unless specified
otherwise all joinery in affected rooms to receive one
coat of undercoat and one coat of gloss as existing.
External rendering to receive primer coat on new
work followed by two coats of resin based masonry
paint on the full
elevation wall on
panel, as indicated
be confirmed to the

8.   PROTECTION OF F
Furniture and
gauge polythene
tape. Carpet edge
vicinity of plaster
are not included
against dusty by se

N.B.   The abo
only abnd may be
items in the sched
depending on the
plaster which is
removed.

The Works

*N.B. Prices inserted against each item should allow for
all elements such as scaffolding, compliance with
Health and Safety Regulations etc, which are essential
to carry out the work broadly as described in the item.
Unless specified otherwise tenderers should assume
that furniture will remain in the property, carpets will
remain in place and curtains, ornaments etc. will be
removed by the occupier. Prices should allow for the
protection of furniture and carpets as necessary in
accordance with clause 8 of the specification and for
the removal and re-fitting of light fittings, radiators etc.
as necessary.*

1.   Repair cracks
locations
(a)   Front elevation
window and
window.
(b)   Gable taking
the dining roo
(c)   Main house re
dining room w
(d)   Offshoot left w
the pantry win
(e)   Offshoot rear v
bedroom wind
the wall and re
(f)   Rear single st

2.   Carry out internal c
and ceiling of the f

(a)   Front lounge
(b)   Dining room
(c)   Rear bathroom
(d)   Pantry
(e)   Offshoot hallw
(f)   Front bedroom
(g)   Rear bedroom
(h)   Hallway and l
(i)   Small stores
(repair cracks

3.   Allow for crack repairs and redecoration of ground floor
kitchen to both walls and ceiling. Allow for re-tiling. The
insured is intending to provide a new kitchen and
localised tiling should be carried out in conjunction with
his requirements.

4.   Rear offshoot bedroom – allow for removal of the plaster
to the rear wall and internal cross wall and replastering
as existing. Carry out crack repairs to the other walls
and redecorate both the walls and the ceiling. Allow for
adjusting the entrance doorway.

5.   Allow for the provision of 2 lateral restraint straps at first
floor level to both the main house front and rear
bedroom, allow for making good to the skirting and any
wall plaster reinstatement as appropriate.

6.   Provide new frames and doors in the following locations,
making good any plaster work in the vicinity.

a.   rear bathroom
b.   front bedroom internal store
c.   kitchen/ rear passageway

7.   Allow for lifting the skirtings and re-leveling the left hand
section of the rear dining room floor (approximately 1.5
metre width of floor adjacent to the gable affected).
Allow for reinstatement of the skirtings on completion.
Note that the owner is intending to removing the gas fire
and the fireplace prior to the works.

8.   Variation plus or minus on item 2 for wall and ceiling
coverings etc. and not in accorance with the
specification above (£....................). Price only required,
do not extend to tender sum.

9.   Clear site, remove all debris and leave clean and tidy
ready for the replacement of curtains, ornaments etc.

TOTAL OF WORKS TO SUMMARY

Maintenance   Chapter 5

Figure 5.13 *Schedule of remedial work*

starting point for adding details to your sketches. Doors, windows, stairs and fitments should be added along with any defects you come across. This work may involve lifting floorboards, partial removal of skirting, panelling or casings and gaining access to the roof space. Figure 5.14 illustrates a typical set of internal survey sketches for an early 1900s built semi-detached house.

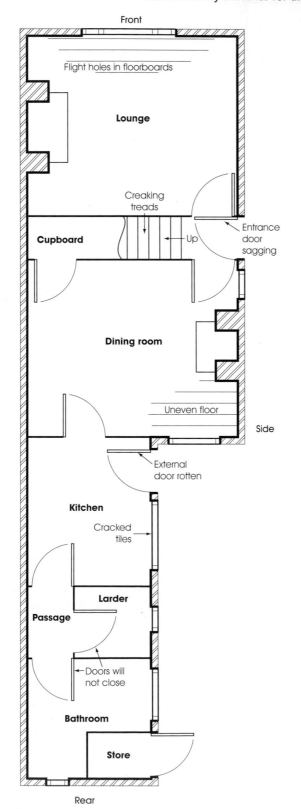

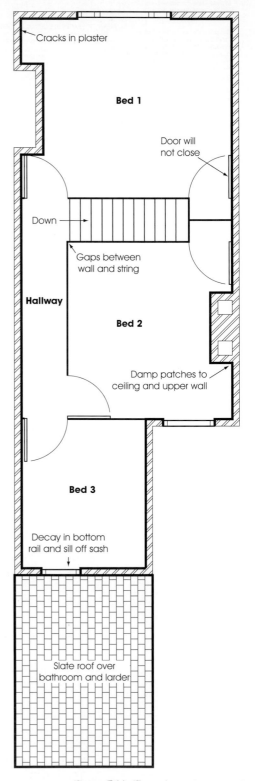

Figure 5.14 *Floor plans showing defects*

Where a lot of detail is required an accompanying list of defects should be made. Table 5.2 shows details of the defects found in the form of a tabled schedule.

Location	Defect	Possible cause	Remedial action
Lounge	♦ Flight holes in floorboards	♦ Woodworm	♦ Expose under-floor space and check remainder of house to determine the extent of attack, then rectify
Dining room	♦ Uneven floor	♦ Possible structural movement (subsidence) or fungal attack	♦ Consult structural engineer ♦ Expose under-floor space to determine extent, then rectify
Lower hall	♦ Entrance door sagging	♦ Joints failed	♦ Dismantle and re-assemble door using external WBP adhesive
Stairs	♦ Creaking treads ♦ Gap between wall and string	♦ Shrinkage between tread and riser, glue blocks loose ♦ Moisture movement and/or failure of fixing	♦ Screw treads to risers. Re-fix glue blocks ♦ Re fix string, cover gap with decorative moulding
Kitchen	♦ External door rotten ♦ Cracked tiling	♦ Wet rot ♦ Movement or accidental	♦ Replace with new door ♦ Replace tiles
Bathroom/ larder	♦ Doors will not close	♦ Door twisted ♦ Hinge bound ♦ Defective ironmongery	♦ Ease rebates or replace door ♦ Scrape off paint, pack out hinge ♦ Adjust or replace tiles
Bedroom 1	♦ Cracks in plaster ♦ Door will not close	♦ Structural movement/ shrinkage ♦ As above	♦ Consult structural engineer ♦ Cut out and repair ♦ As above
Bedroom 2	♦ Damp patches to ceiling and upper wall	♦ Condensation ♦ Defective slates ♦ Deflective flashing	♦ Provide roof space ventilation ♦ Replace ♦ Replace
Bedroom 3	♦ Decayed window	♦ Wet rot due to lack of repainting	♦ Replace window

Table 5.2 *Internal defects schedule*

Where joinery items are to be repaired or replaced, full size details of the sections and mouldings must be made to enable them to be matched up. This task can be eased by the use of a moulding template (see Figure 5.15). The pins of the template are pressed into the contours of the moulding. It can be placed on the sketch pad, drawn around and dimensions added. The location of where the moulding was taken should be noted as they may vary from room to room.

External survey

Sketch outline elevations of the building and add any defects found. Photographs of the elevation may be taken as a backup to your sketches, especially where intricate details have to be replaced. Often defects found on the internal survey are a result of external defects. These should be your starting point. As an example a damp mouldy patch in the corner of a kitchen might be the result of surface condensation, due to poor ventilation. Alternatively it may be due to a leaking rainwater gutter or downpipe. Binoculars are useful for viewing higher levels of a building; closer observation might involve the use of a ladder or the erection of a scaffold. Figure 5.16 illustrates a typical set of external survey sketches used for the inspection of an early 1900s semi-detached house.

Maintenance

Chapter 5

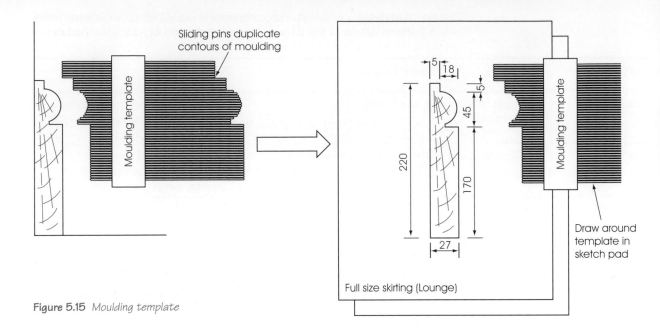

Figure 5.15 *Moulding template*

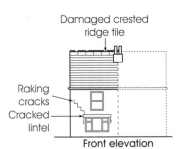

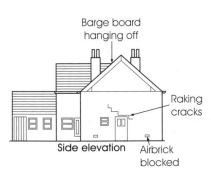

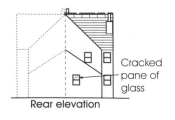

Figure 5.16 *External survey sketches showing defects*

Your focus of attention when carrying out the external survey should include the following points.

Walls

◆ Signs of structural movement: cracks in brickwork joints, rendering and missing pointing.
◆ Staining: particularly just below the roof eaves, above the damp proof course (DPC) and behind rainwater downpipes.
◆ Height of DPC above ground level: this should be a minimum of 150 mm.
◆ Air bricks: ensure they are clear and not blocked by overgrown vegetation.

Windows and doors

◆ Condition of woodwork: look out for poorly fitting doors and casements.
◆ Condition of paintwork: cracked paintwork at joints will allow water penetration and may lead to wet rot. The easy insertion of a penknife or bradawl may confirm your suspicions.
◆ Condition and operation of ironmongery.
◆ Condition of glass, putty and glazing beads.

Roofs

◆ Missing, displaced or damaged tiles, slates and flat roof coverings.
◆ Missing, displaced or damaged flaunching, flashings, valley gutters and verges.
◆ Bargeboards, fascia and soffits: condition of paintwork; look out for signs of decay and distortion. Also check soffits for the presence of any ventilation gaps or grills (these may require cleaning, collected debris or repeated painting may block them).

Guttering, downpipes and drains

◆ Check joints are sealed; look out for signs of water staining and moss growth.
◆ Feel behind cast iron downpipes for corrosion damage due to lack of paint protection.
◆ Check all brackets and clips are secure.
◆ Check drain gullies are clear. Look out for signs of them discharging water over their edges onto a path or house wall.

◆ Check drain gullies are retaining their water seal. If no water is seen in the 'U' bend, it may be cracked or broken and discharging water to undermine the foundations and also causing dampness.

Outside areas

◆ Check condition of garden walls, paths and driveways. Look out for signs of structural movement.
◆ Wooden fences, post and gates: look out for signs of decay. They are particularly susceptible at ground level and joints, where water can be retained.
◆ Note position of trees and other large plants.

Table 5.3 shows the results of the external survey in schedule form.

Table 5.3 *External defects schedule*

Location	Defect	Possible cause	Remedial action
Front elevation	◆ Ridge tile damaged ◆ Cracking to ground floor lintel and raking cracks above	◆ Wind damage/ uncertain ◆ Structural movement	◆ Replace and make good ◆ Consult structural engineer
Side elevation	◆ Air bricks blocked ◆ Raking cracks above entrance door ◆ Barge board hanging off	◆ Build up of dirt, soil and vegetation ◆ Structural movement ◆ Fixings failed	◆ Clean out, reduce ground level to at least 150 mm below DPC ◆ Consult structural engineer ◆ Re-fix and make good slates if required
Rear elevation	◆ Cracked pane of glass to bathroom	◆ Accidental/unknown	◆ Re-glaze window
Outbuildings and structures	(Not viewed)		

Timber repairs

When considering timber repairs a great deal of judgement and negotiation with the client is often required. Can it be repaired cost effectively or is it cheaper in the long run to replace it? Each job has to be considered on its own merits, cost against future service life being the main consideration.

Doors

There are many defects associated with doors. Remedial action will depend on the type and location and may range from a simple adjustment through to complete replacement. Table 5.4 and Figure 5.17 covers the most common defects, causes and recommended remedial action. In all but minor cases consider/discuss with clients the possibility of renewing the door.

Maintenance **Chapter 5**

Table 5.4 *Door defects*

Defect	Sketch	Causes	Remedial action
Will not close, door sticking at head, sill or stile	Stile hits against jamb	◆ Build-up of paint over time or swelling due to intake of moisture	◆ Ease (plane) edge or top ofdoor, remove extra material to allow for clearance; repaint/seal exposed edges on completion
Large gaps between door and frame	Large gap	◆ Shrinkage due to reduction in moisture content	◆ Add lipping to one or preferably both edges of the door as these should be at least 10 mm thick. It may be necessary to further reduce the door width before fitting them; repaint/seal exposed edges on completion
Sagging, tapered gap at top of door may be touching floor		◆ Uneven settlement (downwards movement) of hanging stile, wall and door frame	◆ Remove top bevel, lift updoor to fit frame; add new piece to the bottom of the door
		◆ Loose joints in framed doors	◆ Dismantle door, re-glue joints and reassemble square
Hinge bound or binding door springs or resists as it is being opened and closed	Paint buildup Protuding screwheads Gap Distorted stile Hinge recess too deep	◆ Hinge recess cut too deep.	◆ Pack out hinge recess with a piece of thin card (can be cut from screw box)
		◆ Build-up of paint over hinges	◆ Scrape off paint build-up
		◆ Protruding screw heads	◆ Remove screws; plug holesand replace with smaller headed screws
		◆ Distorted hanging stile	◆ Small distortions can be corrected by adding an extra hinge in the centre, otherwise renew door
Twisted door, does not close evenly to stop	Not closing at bottom	◆ Door distorted often due to irregular grained timber and variable moisture conditions particularly if not completely painted/ sealed	◆ For small gaps, stops can be adjusted or rebates eased to mask situation, otherwise renew door
		◆ Door frame fixed out of line	◆ As for small gaps or re-fix frame in-line

Decayed or damaged stiles, rails or panel.		◆ Absorption of moisture particularly into unsealed end grain and opened joints ◆ Accidental or criminal damage	◆ Cut away decayed or damaged section and repair using keyed splices and falsetenons
Door does not stay latched when closed.		◆ Build-up of paint over time to the stops or rebate ◆ Distorted door ◆ Defective lock/ latch	◆ Ease stops/scrape off paint build up; adjust striking plate ◆ Rectify distortion and adjust striking plate ◆ Replace lock/latch
Damaged panel		◆ Shrinkage; accidental or criminal damage	◆ Cut out panel and stuck (moulded on solid) beads to form new rebate in place of panel groove; fit new panel and secure with planted beads.

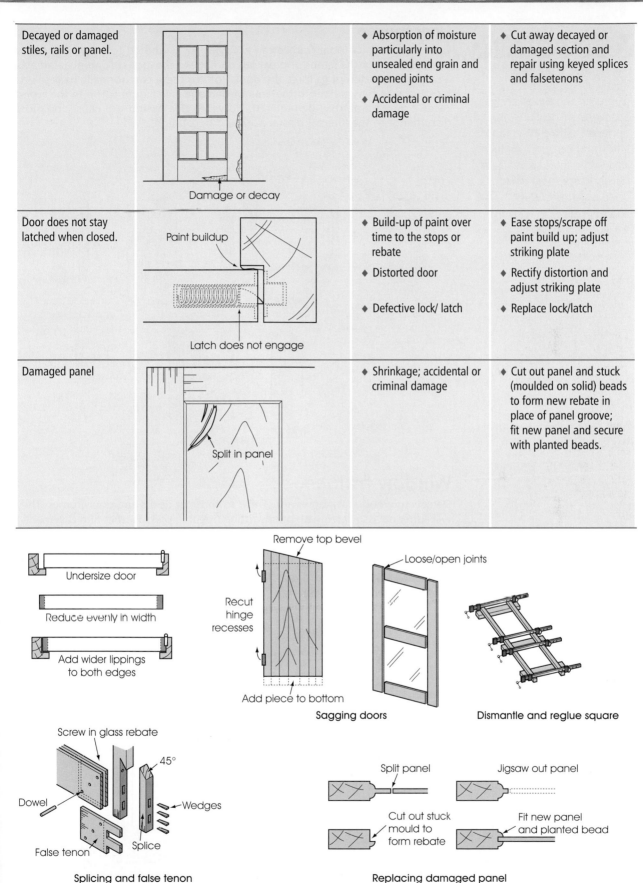

Figure 5.17 *Remedial treatment to doors*

Chapter 5 | Maintenance

Door frames

Defects to external door frames can normally be attributed to either wet rot to the lower end of the jambs or breakout damage in the lock striking plate area as a result of an attempt to force the door. Both of these can normally be resolved by splicing in new timber to the area of damage (see Figure 5.19). In the worst cases a new frame may be required. However, this option will cause the most disturbance to the internal plasterwork and decoration.

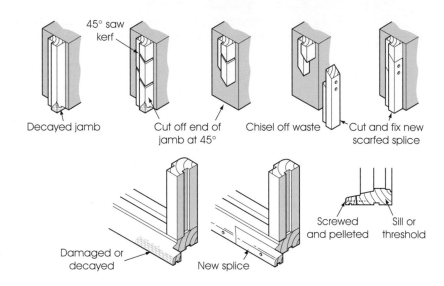

Figure 5.18 *Repairs to door and window frames*

Window frames

Defects to wooden windows are similar to those of doors and door frames. They will either be associated with poorly fitting opening parts (sashes and casements) or decayed/damaged frames. These can be rectified using the methods previously outlined for doors, e.g. scraping off the paint build-up, easing leading edges, dismantling and re-assembly of sagging casements and splicing of new timber to decayed or damaged areas. In the worst cases the installation of a new window should be considered.

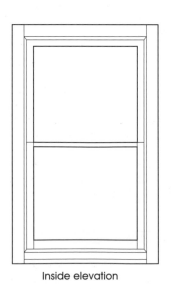

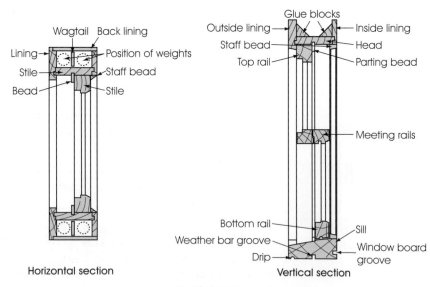

Figure 5.19 *Boxed frame sliding sash window details*

Boxed frame sash windows

This type of window is the traditional pattern of sliding sashes and for many years has been superseded by casements and solid frame sash windows. This was mainly due to the high manufacturing and assembly costs of the large number of component parts. An understanding of their construction and operation is essential as they will be met with frequently in renovation and maintenance work.

The double-hung boxed window consists of two sliding sashes suspended on cords that run over pulleys and are attached to counter-balanced weights inside the boxed frame.

Figure 5.19 shows an elevation, horizontal and vertical section of a boxed frame sliding sash window. It shows the make-up of this type of window and names the component parts.

Re-cording sashes – the maintenance carpenter is often called upon to renew a broken sash cord (Figure 5.20). It is good practice to renew all four cords at the same time, for the remaining old cords will be liable to break in the near future.

The sequence of operations for renewing sash cords is as follows:

◆ Carefully remove staff beads.
◆ Carefully remove pockets (access pieces cut towards the bottom of pulley stiles).
◆ Take out bottom sash. The sash cords should be wedged at the pulley and removed from the groove in the sash.
◆ Carefully remove parting beads. Break paint joints first and carefully prise out with a chisel.
◆ Take out top sash in a similar manner to the bottom sash.
◆ Remove the weights and cords through the pockets. The wagtail will move to one side to give access to the outside weights.
◆ Thread new cords over pulleys and down to the pockets. A 'mouse' can be used to thread the cords easily. A 'mouse' is a small lead weight that is attached to a 2-metre length of string that in turn is tied to the cord. The mouse is inserted over the pulley and drops to the bottom of the frame. The sash cord can now be pulled through. Many carpenters use a length of small chain instead of a mouse.
◆ Fasten sash cords to weights. To obtain the length of cord for the top sash, rest the sash on the sill and mark on the pulley stile the end of the sash cord groove. Pull the weight up to almost the top and cut the cord to the position marked on the pulley stile. Wedge the cord in the pulley to prevent the weight from dropping. To obtain the length of cord for the bottom sash, place the sash up against the head of the frame and mark on the pulley stile the end of the sash cord groove. With the weight just clearing the bottom of the frame cut the cord to the position marked on the pulley stile. Wedge the cord in the pulley.
◆ Fix sash cords to the top sash and insert the sash into the frame. The cords are normally attached to the sashes by nailing them into the cord grooves. Alternatively the cord can pass through a closed groove and end in a knot.
◆ Replace the parting beads. Where these have been damaged new ones should be used.
◆ Fix the sash cords to the bottom sash and insert the sash into the frame.
◆ Replace the staff beads and check the window for ease of operation. Candle wax can be applied to the pulley stiles and beads to assist smooth operation.

Door and window hardware

Incorrect cleaning methods can ruin door and window ironmongery (hardware). It is important to ensure proper care is taken to keep them clean. Dust, contamination and moisture are the main hazards that can affect door

did you know?

The weights in sash windows may not all be the same, so ensure they are returned to their original positions.

Maintenance Chapter 5

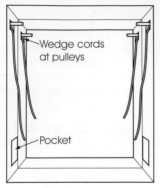

Working from inside,
wedge cords and
remove sashes

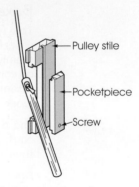

Remove cords and
weights through pockets

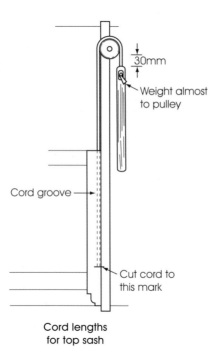

Cord lengths
for top sash

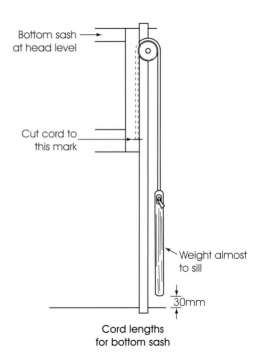

Cord lengths
for bottom sash

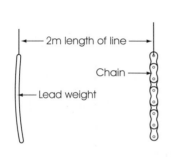

Types of mouse

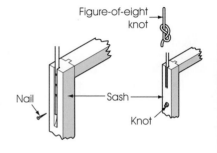

Cord nailed or knotted

Figure 5.20 *Re-cording sashes*

and window hardware. Irreparable damage can be caused to the surface by using proprietary metal polishes, harsh abrasive cleaners or emery cloths on electrolytically deposited or special surface finishes. The correct cleaning procedures for the most common finishes are as follows:

◆ Anodized aluminium finishes should be dusted regularly, washed periodically with weak detergent solutions and occasionally wiped with wax polish.
◆ Stainless steel finishes should be dusted regularly and occasionally washed with soap and water.
◆ Plastic products require only wiping with a damp cloth. Strong sunlight can cause surface deterioration that cannot be easily rectified.
◆ Powder coated finishes should be cleaned with a soft cloth and household furniture polish. Industrial solvents must not be used as these can destroy the finish.
◆ Lacquered finishes (polished brass etc.) should be cleaned occasionally with good quality wax-free polish. Abrasives and metal polishes should not be used as these will remove the lacquer.

The moving parts of locks, latches, bolts and hinges require regular lubrication and need to be kept free from paint build-up, in order to ensure their trouble-free operation.

The following tips can be used when maintaining door and window hardware:

◆ Always follow the specific manufacturer's instructions when checking and maintaining hardware.
◆ Check for the ease of operation of all moving parts.
◆ Ensure all items are kept free from paint build-up.
◆ Check that all screws have been fitted, ensure that they are tight and of the correct size.
◆ Check that hinges are in true alignment and the recesses are equal in depth, square and plumb.
◆ Lubricate the knuckle joints of metal hinges three times a year.

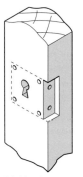

Old lock removed

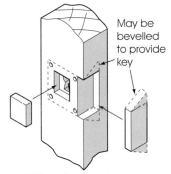

May be bevelled to provide key

Filling pieces cut, holes and recesses enlarged

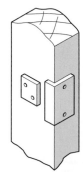

Oversize filling pieces glued and pinned in place

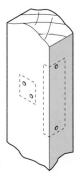

Filling pieces planed off flush, pin holes filled and sanded off flush

Figure 5.21 *Making good holes and recesses*

◆ Check the alignment of lever furniture, striking plates and keyhole plates.
◆ Check that the return springing on lever furniture and door knobs is satisfactory.
◆ Apply graphite to keys and repeatedly insert them into the lock, if the key or lock has a tendency to stick or is stiff.
◆ Check the mortising of lock cases and striking plates, if the latch or deadbolt sticks or will not engage. Graphite can also be applied to the leading edge of a latch bolt to assist its movement into the striking plate.

Replacement of hardware

Inevitably they will eventually begin to wear and require replacement. In general these should be replaced with like for like or the nearest alternative. However proposed replacement does provide the opportunity to upgrade hardware; particularly locks for increased security.

When replacing or upgrading hardware it is often necessary to make good holes and recesses. Figure 5.21 illustrates a typical situation.

In general all follow the same procedure:

◆ Cut oversize filling pieces of a similar material. These are often bevelled to provide a key.
◆ Mark and cut out enlarged hole and recesses to receive filling pieces.
◆ Glue and pin filling pieces in place.
◆ Punch in pins, plane off flush, fill pinholes with wood filler, sand off smooth and repaint or seal the area.

Frame replacement

When extensive repairs are required it is often more cost effective in the long run to consider a complete replacement. This should be discussed with the building owner. It should be emphasized that although the replacement may be more expensive than the repair initially, the replacement would last a lot longer and also provide the opportunity of upgrading fittings etc. and modification to suit the owner's requirements.

Before removing frames, always check the wall above for signs of support (see Figure 5.22). Newer buildings may have a concrete or steel lintel, older properties a stone lintel, brick arch or soldier course, or sometimes none at all apart from the frame. Also look out for cracks in the brickwork mortar joints above the opening, which could be indications of structural movement, possibly leading to collapse on removing the frame.

Figure 5.22 *Means of support over openings*

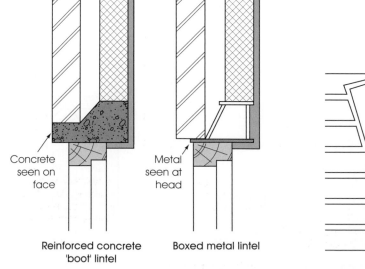

Concrete seen on face

Metal seen at head

Reinforced concrete 'boot' lintel

Boxed metal lintel

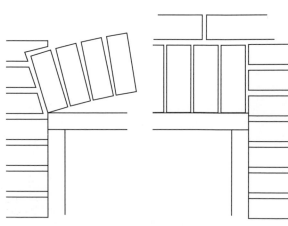

Brick arch or soldier course seen on face

Do not proceed if a means of support is not evident or there are signs of movement. Seek the advice of a structural expert, as arrangements may have to be made for temporary support, the insertion of a lintel and structural repairs to be carried out by others before the frame replacement itself.

Once you are sure the opening is correctly supported, the frame can be cut out and removed in sections in sequence as shown in Figure 5.23.

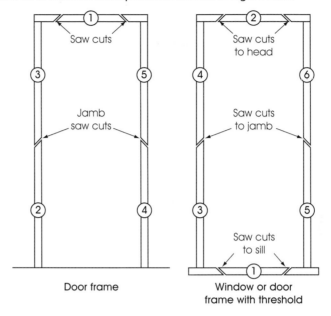

Figure 5.23 *Removing a door frame*

Cut out and remove in sections using numbered sequence

When the frame has been removed, you should clean off any projecting mortar and previous fixings. Make good any damaged brickwork and holes previously occupied by 'built-in' horn fixings. Finally, fix in position the new frame (see Figure 5.24).

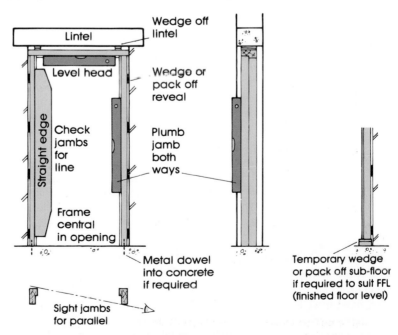

Positioning frame

Figure 5.24 *Positioning and fixing a frame*

Maintenance Chapter 5

◆ Cut off horns and seal cut ends (paint or preservative).
◆ Place frame in the opening, temporarily holding with wedges at head or sill if required.
◆ Check head and sill with a level and adjust wedges as required.
◆ Check jambs for line with a straight edge, plumb one jamb with level, 'sight-in' the other, adjust position and wedge as required.

Floors and roofs

If the defect is the result of fungal decay or wood-boring insect damage, the procedure outlined previously should be adopted. Defective joists and rafters in ground, upper floors and roofs can be repaired on a localized basis by cutting away the affected timber and bolting a new piece of timber alongside. Any new member inserted in this way should normally extend to another load-bearing member, such as a wall plate, binder or purlin etc., as illustrated in Figure 5.25.

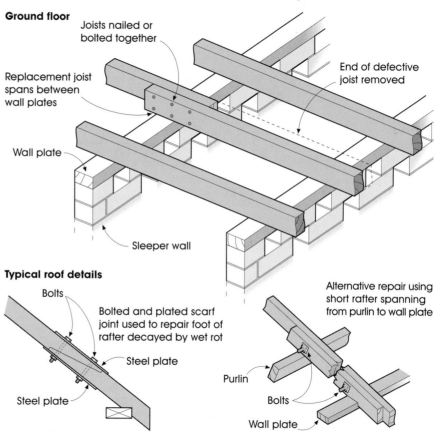

Figure 5.25 *Replacing defective structural members*

However, where the work is extensive, especially in circumstances where structural timber is affected, like upper floor joists, rafters and purlins, etc., it is often wiser, safer and more cost-effective to have this work referred to a specialist contractor who will be fully equipped and experienced to undertake it with confidence.

Floorboards

Again where the defect is the result of fungal attack or wood-boring insects, the procedure explained before should be adopted. If due to movement (loose fixings), wear or damage, one or more boards can simply be re-fixed or replaced.

The surface of the area to be worked on can be scanned before starting work, with a metal/live electric circuit detector. As an added precaution it is wise to turn off all service supplies at their meter/stop valve before any re-fixing or cutting out operations commence.

safety tip

Care must be taken when re-fixing or replacing floorboards as there is always the possibility of services (water, gas and electricity) running below.

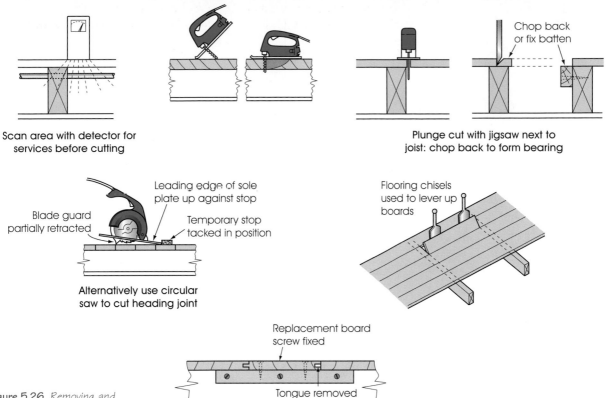

Scan area with detector for services before cutting

Plunge cut with jigsaw next to joist: chop back to form bearing

Chop back or fix batten

Leading edge of sole plate up against stop

Blade guard partially retracted

Temporary stop tacked in position

Alternatively use circular saw to cut heading joint

Flooring chisels used to lever up boards

Replacement board screw fixed

Tongue removed

Figure 5.26 *Removing and replacing floorboards*

Maintenance Chapter 5

Loose boards are best re-fixed by screwing down into the joists rather than nailing, especially at upper floor levels in older properties, where significant amounts of nailing causing vibration may damage the lath and plaster ceiling.

Removing a floorboard (Figure 5.26)

◆ A jigsaw can be used to cut the ends of a floorboard next to a joist. The ends can then be chopped back to form a bearing (for heading joints) using a wood chisel, or fix a batten to the edge of the joist.
◆ Alternatively a small circular saw may be used to cut the heading joints over the joists. The blade should be set to the floorboard thickness.
◆ A sharp knife, padsaw or jigsaw can be used along the length of the board to separate the tongue.
◆ Punch the fixing nails through the board. Insert two wide blade flooring chisels in one of the edge joints and lever up the board.
◆ Where more than one adjacent boards are to be replaced the heading joints should be staggered over different joists.
◆ Finally, cut the new boards to length and re-fix in place. Where more than one board wide a folding technique should be used.

Stairs

The most common defects encountered are illustrated in Figure 5.27.

◆ *Creaking treads* – this results from movement between the tread and riser joint when walked on. Where access to the underside of the flight is possible (in cases where it has not been plastered over) the creaking can be remedied by renewing the glue blocks at the tread to riser junction and re-gluing/re-wedging the treads and risers in their string housings. Where access to the underside is not possible, the problem can be remedied by gluing and screwing the tread down into the top of the riser. Small gaps, normally the result of shrinkage, may be apparent at the tread-to-string housing. These should be filled by gluing in thin strips of timber veneer.

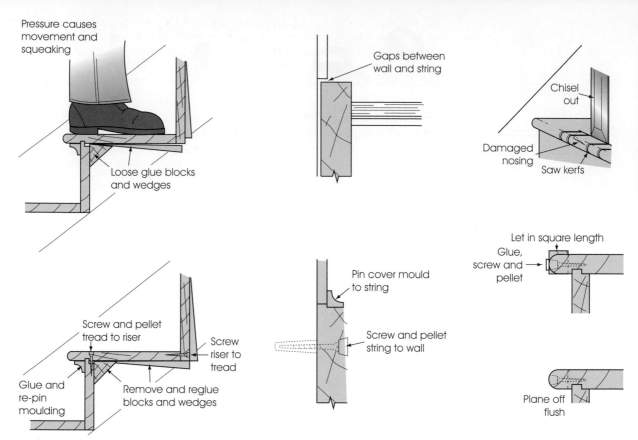

Figure 5.27 *Common stair defects*

◆ **Gaps between the wall and string** – this may be the result of either shrinkage or movement, or a combination of both. Shrinkage gaps can be masked by the application of a cover mould. Movement gaps are normally the result of the fixings between the wall and string becoming loose or failing, causing the string to move away from the wall when using the stairs. Re-fix the string back to the wall by plugging and screwing. Any gap or damage between the string and plasterwork can again be masked by a cover mould.

◆ **Damaged nosings** – these can be repaired by cutting out the damaged section and splicing in a new piece. A square length should be 'let-in'. Cut at 45° at either end for additional support and glue line. This is fixed by gluing and screwing. Finally plane and rub down to match the nosing profile. Where a scotia mould is used at the underside of the tread to riser junction, it is best to renew the whole length, gluing and pinning the new one in place.

Brickwork and plastering repairs

You may be required to replace one or two bricks that have been removed or damaged during other work, or even lay several courses to fill in, say under the sill of a reduced height replacement window.

Replacing bricks

The first task is to attempt to match the pattern, size and make of the original. Measure the brick size and take a small piece to the brick supplier for them to

identify. Brick sizes vary slightly due to the way they are made. New metric-size bricks are a little smaller than the old imperial ones. When working on older property matching bricks may be difficult to obtain. Try suppliers that specialize in reclaimed building materials. If imperial bricks cannot be obtained, matching metric ones can be bonded into the existing work by slightly increasing the mortar joints.

Figure 5.28 illustrates the procedure to follow when replacing a single brick. In this case the brick previously cut around is the built-in horn of a sill or threshold.

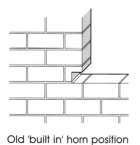

Old 'built in' horn position

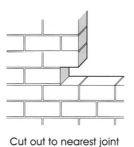

Cut out to nearest joint

Cut replacement brick to size

Lay bed joint, 'butter' up end and top edge of brick

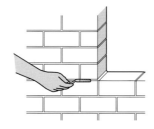

Place brick in position and clean off excess mortar

Rake out adjacent mortar joints and point up to match existing pointing

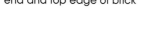

Weather pointed

Tooled (bucket handle)

Recess

Figure 5.28 *Replacing a brick*

- Using a bolster and club hammer (do not forget to wear eye protection goggles), cut out the old brick and clean away the old mortar joints. Brush out all dust particles, taking care in modern buildings to ensure nothing is allowed to enter the cavity.
- Cut the brick to size if required. First gently score a line around the brick using a bolster and club hammer; finally use a heavier blow on the lines to sever it.
- Prepare a mortar mix, typically 1:6 (one part Portland cement to six parts bricklaying sand). Sufficient water is added so that the mix has the consistency of soft butter (firm enough not to collapse when heaped, but easily compressed with a shovel). One part lime or a mortar plasticizer may be included in the mix for improved workability. Alternatively, pre-packed bricklaying mortar mixes are available.
- Lay the bed of mortar into the prepared hole.
- Butter up (apply mortar) to the end and the top edge of the brick. Place it in position and then remove the surplus mortar.
- Fill any gaps in the mortar joint. When it has started to go off, rake the joints out below the brick surface.
- After about 24 hours rake out the adjacent mortar joints.
- Re-point the joints with a mortar mix that matches the original in colour and finish.

Maintenance

Chapter 5

Repointing

The most common methods of repointing are:

◆ weather pointed, which is done with a pointing trowel
◆ tooled, a concave finish (known as bucket handle) created by working along the drying mortar with a special jointing tool, or alternatively a metal bucket handle (hence the name) or a piece of 15 mm copper pipe may be used
◆ recessed, created by brushing out the drying mortar with a stiff bristle hand brush, alternatively a more consistent depth can be achieved by working the joint with a piece of timber having a protruding countersunk head screw.

Relaying brick courses

Figure 5.29 illustrates the procedure to follow when relaying or building whole courses of bricks. This example refers to under the sill of a reduced replacement window.

◆ Cut out bricks at either end to form the bond between the adjacent vertical joints. Clean away any old mortar and dust particles.
◆ Dry lay the bricks to determine the pattern. Cut the bricks to size if required.
◆ Prepare a mortar mix as before, using a 10 mm joint. Approximately 1 kg of mix is required for each brick.
◆ Apply a bed of mortar to the existing brick course, approximately 10 mm thick. Furrow the surface to a 'V' shaped groove with the point of a trowel. Butter up and lay bricks, removing surplus mortar as you go.
◆ Repeat the process to lay the subsequent brick courses. Periodically check the bricks are being laid horizontal and in line with a spirit level. Use the end of the trowel to tap the bricks into place if required.
◆ Complete the job by raking out the mortar joints and finally pointing them as before.

Repairs to plasterwork

Modern buildings – the internal brick and blockwork walls will have a hard plastered finish. This is normally applied in two layers, a 9 to 12 mm thick backing coat and a 2 to 3 mm thick finishing coat. Ceilings and stud partition walls are surfaced with sheets of plasterboard. These may be finished by a 2 to 3 mm coat of board finishing plaster applied on to the plasterboard, which acts as the backing. Alternatively the plasterboard joints may be taped up and filled to provide a 'dry-lined' finish, ready for decoration. Ceilings that were 'dry lined' were often decorated using a textured coating, worked to create a repeating pattern or stipple finish.

Older buildings – the wall plaster may be much softer. This is still normally two coats, namely a thick, lime-based backing coat followed by a thin finishing coat. Ceilings and stud partition walls were then finished using 'lath and plaster'.

This is a system using thin timber laths nailed to the undersides of joists and faces of studs. Wet plaster was pressed up against them and allowed to squeeze between the gaps in the laths forming a key to hold this backing plaster in place. This was finally finished using again a thin coat of finishing plaster.

External walls

The external walls of both modern and older buildings may be covered in rendering. This is a surface coat of sand and cement mortar applied to a wall for decorative and/or waterproofing purposes.

All of these finishes may crack due to structural movement or damage by accidental impact or be disturbed during renovation work, therefore requiring repairs that the maintenance carpenter and joiner may be asked to undertake.

safety tip

The use of barrier cream is recommended when undertaking bricklaying, plastering and tiling. This prevents the chemicals from attacking and drying out the skin.

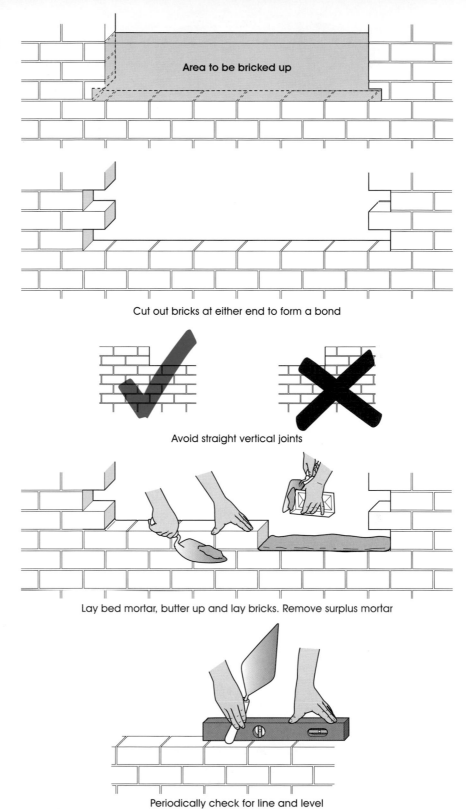

Area to be bricked up

Cut out bricks at either end to form a bond

Avoid straight vertical joints

Lay bed mortar, butter up and lay bricks. Remove surplus mortar

Periodically check for line and level

Figure 5.29 *Laying brick courses*

Patching plasterwork

The first thing to do when patching plasterwork on any background is to protect the floor by covering with a dustsheet. Figure 5.30 shows the procedure to follow when patching a 'blown' or damaged area of plasterwork to a brick or blockwork wall.

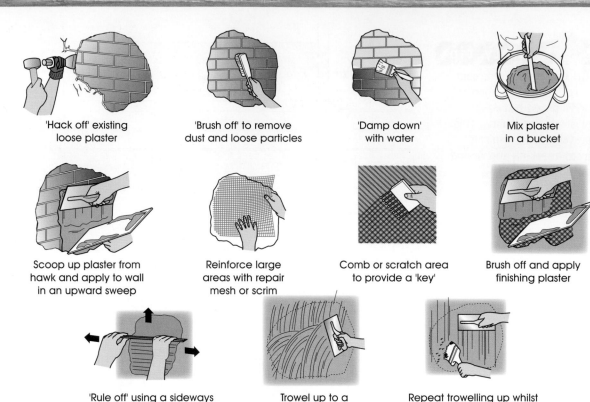

'Hack off' existing loose plaster

'Brush off' to remove dust and loose particles

'Damp down' with water

Mix plaster in a bucket

Scoop up plaster from hawk and apply to wall in an upward sweep

Reinforce large areas with repair mesh or scrim

Comb or scratch area to provide a 'key'

Brush off and apply finishing plaster

'Rule off' using a sideways sawing action, working upwards

Trowel up to a smooth finish

Repeat trowelling up whilst splashing with water

Figure 5.30 *Patching plasterwork*

◆ Tap plaster around the damaged area to 'see' (hear) if any part sounds hollow or loose.

◆ Use a club hammer and bolster to hack off all existing loose plaster until the surface is sound.

◆ Brush down the surface to remove all loose particles and dust, using a stiff bristle or wire brush.

◆ Damp down the wall surface by brushing or spraying with water. This prevents the wall suction from drying out the plaster too quickly, which could result in cracking on drying. Some surfaces such as concrete, shiny or glazed bricks and impervious engineering bricks do not help the plaster to stick. In these cases brush on a PVC bonding agent before plastering to ensure good adhesion.

◆ Add a small amount of backing plaster into clean cold water in a bucket. Stir with a timber stick until a thick, creamy consistency is achieved.

◆ Transfer some of the plaster to your hawk. With the hawk tilted away from the wall scoop up a small amount of plaster on the edge of the steel trowel and press the plaster against the wall using an upward sweep of the trowel. The trowel should be used at an angle to the wall, with the angle reduced as you sweep it up the wall. Take care not to allow the trowel to lay flat against the wall, as the suction will pull the fresh plaster off the wall.

◆ Continue adding plaster to the wall until the whole area to be repaired is covered, to within 2 to 3 mm below the surrounding wall finish. Larger areas may be reinforced by pushing a repair mesh or scrim into the backing coat.

◆ Before the backing plaster is completely set, scratch the surface with a comb. This provides a 'key' to help with the adhesion of the finishing coat.

◆ After about three to four hours the backing coat surface will be hard, but not dry. It is then ready for finishing. If allowed to dry further, it will require damping down again with water before finishing.

◆ Brush down the surface to remove any loose particles. Mix up a small batch of finishing plaster, by adding to a little water as before, except this time it should be of a runnier consistency. Trowel on the plaster, aiming to leave it slightly proud of the surrounding area.

◆ Rule flat the surface, using a timber or metal straight edge. Start at the bottom of the patch, move it up the wall with a side-to-side 'sawing' action, keeping it tight against the existing sound plaster as a guide. Trowel on more plaster to fill any hollows before ruling off again.

◆ The plaster will start to set within 30 to 45 minutes. At this stage smooth the surface with a plastering trowel. Again using upward sweeps with the trowel held at an angle. After 15 or so minutes lightly splash the surface with clean cold water, while trowelling up and over the surface, to provide a smooth hard finish. Ensure the trowel is kept clean and damp during this process, to prevent damaging the newly plastered surface. Regular brushing off in a bucket of cold water is ideal.

Note: Small patches in rendering can be repaired in one or two coats, using the above procedure, except that a sand and cement mix is used in place of plaster.

Patching damage to plasterboard or lath and plaster surfaces – to repair small holes caused for example by striking the surface with the corner of a piece of furniture when moving it. The procedure to follow is illustrated in Figure 5.31.

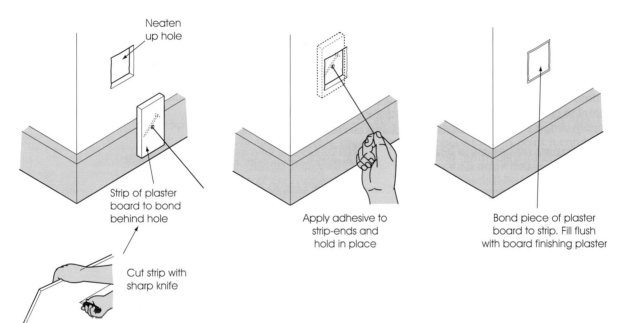

Figure 5.31 *Patching small holes in plasterboard*

◆ Neaten up the jagged edges using a sharp knife for plasterboard or a padsaw for lath and plaster.

◆ Cut a strip of plasterboard about one-and-a-half times the length of the neatened hole and just narrower than it in width. Make a hole in its centre, pass through a piece of string and knot it behind on a nail.

Note: Plasterboard is simply cut by scoring on the face with a sharp knife, breaking along the line by applying pressure along from the scored side. Run the knife along the paper on the other side to separate.

◆ Mix up some plasterboard adhesive and apply to both ends of the strip. Feed the strip into the hole, using the string to pull it tight against the inner face of the board or laths. Tie off the string to a scrap of timber positioned over the face of the hole.

◆ When the adhesive has set, cut off the string. Cut another piece of plasterboard, this time to fit the hole and again bond in place using plasterboard adhesive. Press in until it is just below the surrounding wall surface, leave to set.

◆ Fill the patched hole using board finishing plaster and trowel up as before.

Maintenance

Chapter 5

To repair large holes, for example caused by a foot slipping through the ceiling when working in a loft, the procedure to follow is illustrated in Figure 5.32.

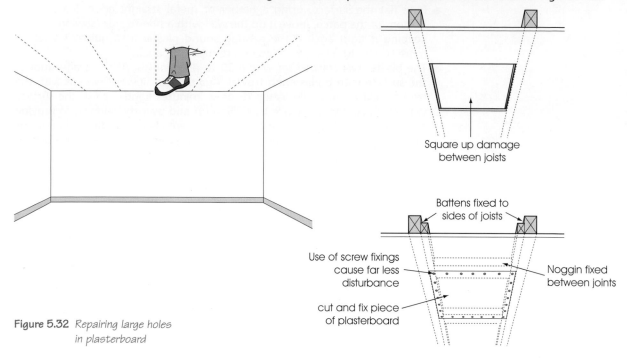

Figure 5.32 *Repairing large holes in plasterboard*

◆ Mark on the ceiling two lines at right angles to the joist direction and enclosing the damaged area.
◆ Use a padsaw to cut along these lines until the adjacent joists are reached after first checking for the presence of cables and plumbing.
◆ Again using the padsaw, cut along the joist edges. Remove the damaged area, leaving a neat rectangular hole.
◆ Cut battens; fix to the sides of the joists.
◆ Cut and fix noggins between the battens at either end of the hole, ensuring the noggins centre lines straddle the cut line, to provide a bearing for both the existing sound ceiling and new plasterboard.
◆ Cut a piece of plasterboard 2 to 3 mm smaller than the hole in both directions. Fix in place using plasterboard nails or plasterboard screws into the noggins and battens.
◆ Use a sharp knife to cut away the finishing plaster about 25 mm all round the hole. Bed lengths of plasterer's scrim over the joints between the patch and existing sound ceiling, using a thin (runny) mix of board finishing plaster as an adhesive. This is to reinforce the joint and reduce the risk of later cracking.
◆ Fill the patched area using board finishing plaster and trowel up as before.

To make good the plasterwork around the reveals of a replaced door or window: in most circumstances these can be patched using the two coat backing and finishing plaster method as before. Where there is extensive damage, or the plaster is loose, the entire reveal should be hacked off and replaced. Figure 5.33 illustrates the procedure to follow using plasterboard and board finishing plaster:

◆ Hack off plaster reveal back to the brick or blockwork surface and extending around the corner by about 75 to 100 mm.
◆ Cut two strips of plasterboard, one for the reveal and the other for the return.
◆ Brush down the wall surfaces. Mix up some plasterboard adhesive. Using a trowel or special caulker, apply dabs of adhesive up the centre of the reveal and continuously around the perimeter.

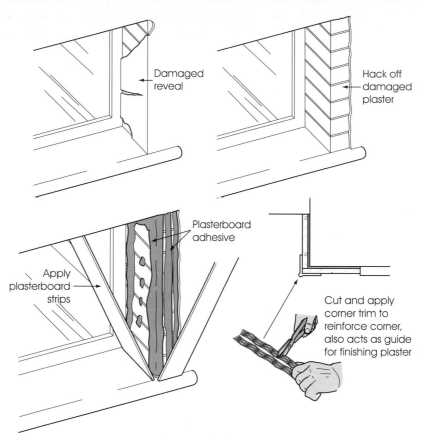

Figure 5.33 *Making good plasterwork to damaged reveals*

◆ Press plasterboard strips in place and check for plumb with a spirit level. The return strip should finish 2 to 3 mm below the adjacent plaster, to allow for a coat of board finish.

◆ Reinforce the corner with a length of metal plasterboard bead (this also acts as a guide for the later board finishing plaster coat). Apply a bed of plasterboard adhesive or board finish to the corner. Press the bead in place, check with spirit level for plumb. Also ensure it is in line with existing wall surface. Remove excess adhesive and allow to dry for two to three hours.

◆ Reinforce the joint between return plasterboard strip and existing wall plaster, using a length of scrim as before.

◆ Complete repair by applying a coat of board finishing plaster and trowel up in the normal way.

Replacing damaged tiles

You may be required to replace damaged tiles individually or lay a much larger area such as a whole wall or floor.

Probably the hardest task is to find a good match for replacement. The building owner may have spares left over from the original work. Alternatively, take a damaged piece to a tile supplier for them to find a match. Figure 5.34 illustrates the procedure to follow.

◆ Prepare the surrounding area and yourself. Cover the floor and surrounding units or bath and sanitary ware with dustsheets to protect from dust and possible scratching by the small, sharp particles of broken tiles. Protect yourself by wearing eye protection goggles, gloves and a dust mask.

◆ Rake out grout joint around damaged tile, to relieve the perimeter stresses.

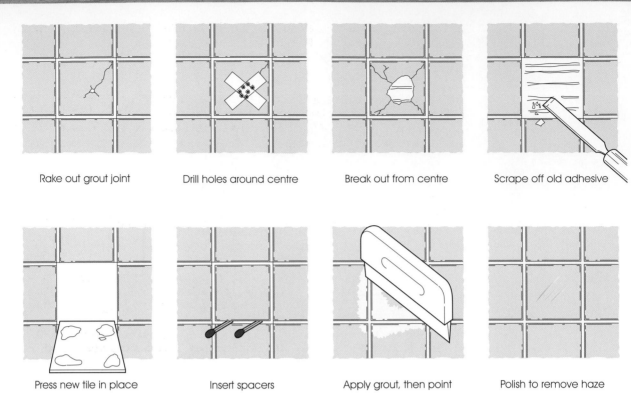

Rake out grout joint   Drill holes around centre   Break out from centre   Scrape off old adhesive

Press new tile in place   Insert spacers   Apply grout, then point   Polish to remove haze

Figure 5.34 *Replacing a damaged tile*

did you know?

The long-term success of tiling depends on the adhesive and grout used. Most are available either in ready mixed or powder form for mixing with water to a creamy paste with water. A standard-type mix is only suitable for dry areas. It may also tolerate a little condensation or occasional splashing. In areas subject to more prolonged condensation or extensive wetting such as a shower area, always use waterproof products.

◆ Drill a series of holes around the centre of the tile and break out using an old chisel, working progressively towards the edges. Do not try to break out the tile by trying to prise it off from the edge joints, as almost inevitably you will damage the adjacent tiles. Masking tape can be applied when drilling out the centre, to prevent the masonry drill skidding across the ceramic surface.

◆ Scrape or chip off the old tile adhesive back to the surface, taking care not to damage the plaster base.

◆ Apply four dabs of tile adhesive to the back of the tile, or use a notched comb to provide a uniform ribbed layer.

◆ Press the tile in place, so that it lies flush with the surrounding tiles. Insert tile spacers or matchsticks in the joints to position or support the tile. Adjust the tile as required to ensure a uniform gap all around.

◆ Allow the adhesive to set for about 24 hours and then remove the spacers.

◆ Fill the gap around the tile with grout, working in with a rubber squeegee. Point the joint with a finger tip or piece of wood dowel with a rounded end point.

◆ Finally when dry polish up the surface to remove the grout haze, using a clean dry cloth.

Cutting tiles

Tiles are best cut using a proprietary tile cutter, or diamond tipped wet saw. However, small amounts can be cut by scoring the surface with a carbon tipped tile cutting point and snapping the tile along the scored line. Place two matchsticks under scored line. Press down firmly on either side to snap in two (see Figure 5.35). Corners and curves can be cut out of tiles to fit around projections using a tile saw blade in a coping saw frame. Alternatively the lines of the area to be removed may be scored and the waste nibbled away with a pair of pincers.

Grouting

When all cut tiling is complete, allow to set for about 24 hours. The joints between them can then be grouted. Working grout with a rubber squeegee,

point up with finger or a rounded end dowel. Allow to dry and polish off the haze with a dry clean cloth as before.

Finishing off

The joint between tiles and horizontal surfaces such as kitchen worktops, baths and sanitary ware will require sealing with a silicone sealant to prevent moisture penetration. Figure 5.36 illustrates the procedure to follow, which ensures a neat bead of sealant to these locations.

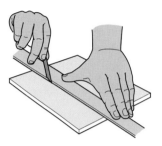

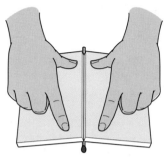

Figure 5.35 *Cutting ceramic tiles*

Score along line with tile cutting point

Place matchsticks under score, press down to snap

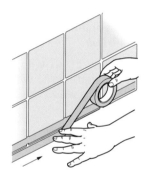

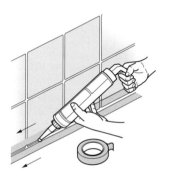

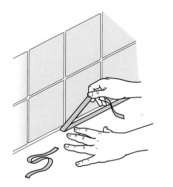

Apply masking tape along angle to be sealed

Push the cartridge nozzle along the angle

Allow sealant to skin over, peel off tape

Complete set bead

Figure 5.36 *Sealing tiles to a worktop*

Apply masking tape to both the vertical and horizontal surfaces, 2 to 3 mm away from the internal angle. Trim the cartridge nozzle off at an angle of 45° to give a bead just wide enough to fill the gap between the two taped edges. Gently squeeze the cartridge gun trigger until the sealant is just seen at the tip. Place the nozzle at one end of the angle to be filled, with the gun held at 45° to the wall. Apply steady, even pressure to the trigger while pushing the gun along the angle. The nozzle will form the sealant into a neat concave curve. To stop the flow at the corners and on completion, release the metal tag adjacent to the trigger. Any unevenness in the bead can be smoothed using a small paintbrush dipped in water. Leave sealant for a short while to skin over; then peel off the tape to leave a well-formed neat bead.

Maintenance **Chapter 5**

Glazing and painting

Maintenance of glazing work involving sealed double-glazing units and large window panes is best undertaken by specialist glaziers, who will be kitted up to undertake the work efficiently and safely.

Re-glazing

Re-glazing of small single-glazed panes can be undertaken by the carpenter and joiner. Wherever possible, glass should be pre-cut to size by the glass supplier. Measure the timber rebate sizes and order glass 3 mm undersize in both directions. For example, a piece of glass for a rebate opening size of 150 × 250 mm should be ordered as a cut size of 147 × 247 mm. If necessary pieces of glass can be cut to size by scoring along the required line using a glass-cutting wheel. Lay the glass on a flat surface with matchsticks placed under the scored line at either end. Apply pressure on both sides to snap in two along the line. Never attempt to cut narrow strips and always wear eye protection and gauntlets.

The first stage is to remove the broken pane. Where practical, sash or casements should be removed from their frame so that broken glass can be removed and replaced with a new piece with relative safety, at ground level.

When replacing broken glass at high level ensure that the area below is cordoned off so that no one can enter the area below. Before starting to hack out the broken pane and hardened putty, ensure you are wearing eye protection goggles and gauntlets to protect hands, wrists and lower arms. These should be worn during the whole process as inevitably shards of glass and fine splinters will be created as the pane is removed and the rebates cleaned up.

Start at the top of the pane removing the old putty or glazing beads with a wood chisel or hacking knife. Remove glazing sprigs (flat or square nails) with pliers and then lever out remaining glass from behind again using a wood chisel. Finally, continue hacking out remaining back putty to rebates.

The procedure for re-glazing is illustrated in Figure 5.37.

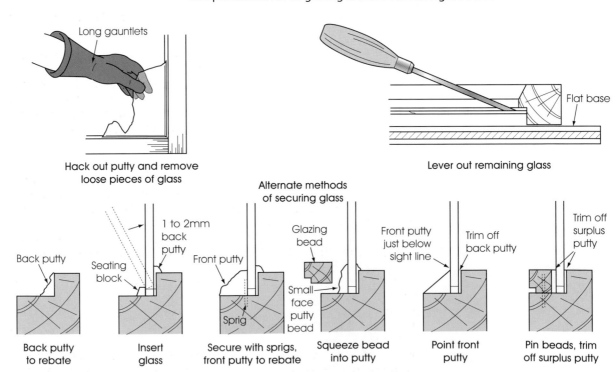

Figure 5.37 *Re-glazing procedure*

◆ Check glass is correct size.
◆ Ensure rebates are primed with paint.
◆ Work a bead of putty around the back of the rebate.
◆ Position plastic seating blocks in the bottom of the rebate to support the glass.
◆ Position the bottom edge of the glass on seating blocks. Gently push the glass into the rebate, applying pressure evenly around the edge until a back bed putty thickness of 1 to 2 mm is achieved.
◆ Use glazing sprigs to secure the glass in place. These may be driven in using a pin hammer or the edge of a firmer chisel. (Panel pins should not be used to secure the glass in place, as their round point of contact results in pressure points, which can lead to the formation of cracks.)
◆ Work a bead of putty all around the rebate in front of the glass. Only a small bead is required when using glazing beads to secure the glass.
◆ Replace glazing beads squeezing putty into glass, or point up the front putty bead using a putty knife or chisel to form a bevelled fillet. Its upper edge in contact with the glass should be just below the rebate sight line. Slight imperfections in the putty fillet can be improved by running over with a wetted paintbrush.
◆ Trim off the surplus putty to beads and then trim off the back bedding putty on the inside of the glass.
◆ Clean the glass to remove oil and putty marks before it dries.

Painting woodwork

Woodwork is painted to provide a decorative finish. However, more importantly, it also serves to protect it from the elements. As a carpenter and joiner carrying out maintenance work you may be required to paint new replaced items as well as repaint existing items that have been eased or repaired. The paint system for wood normally consists of a primer to seal the surface and provide a bond for later coats, an undercoat that provides a smooth opaque covering coat and finally a decorative gloss or satin top coat.

Preparation

The key to a successful paint system is careful preparation. Paint will not last long on a defective surface.

Bare wood – should only need an initial rub down with glass paper to remove any roughness and sharp arrises. Knots in bare soft wood can be full of resin and may later 'bleed' through to the finished paint surface if not sealed. Firstly wipe over the surface with a cloth soaked in white spirit to remove any stickiness and excess resin. Then coat all knots with a knotting solution. On resinous hardwoods that are to be painted, wipe off excess resin using white spirit and seal the entire surface with an aluminium wood primer.

◆ Apply wood primer to all surfaces and edges taking particular care to achieve full penetration of any end grain. This is best undertaken prior to fixing in order to ensure full protection. For example, the backs of skirtings, architraves, door frames and linings etc. are inaccessible when fixed. This is particularly important for external timber.
◆ Fill any defects and open end grain with a wood filler. Always select a waterproof type for external use. Rub down flush with the surface using glass paper.
◆ As priming tends to raise the grain of woodwork resulting in a felt-like hairy surface, the whole job will require rubbing down (de-nibbing) prior to over-painting.
◆ Wipe off surface with a 'low tack' cloth to remove any surface dust; apply undercoat.
◆ When undercoat is dry, apply the topcoat. Refer to paint manufacturer's information with regard to minimum and maximum over-coating times. If the top coat is applied too soon, the undercoat will tend to bleed into

Maintenance — Chapter 5

safety tip Always wear eye protection and long gauntlets when undertaking re-glazing work.

safety tip Remember that surfaces painted prior to the 1960s may contain harmful lead in the paint. In these cases it's best to rub down using a wet process, with wet and dry paper to minimize dust. In all circumstances you should wear a dust mask when rubbing down and eye protection when scraping off paintwork.

the top coat causing defects; too long and it may not bond successfully to the undercoat, resulting in early breakdown and peeling off.

Previously painted wood – if in good condition, lightly rub down with glass paper or clean off using a sugar soap solution. This cleans the surface dirt or grease deposits and removes some of the gloss. New coats of paint will not key well on a gloss surface and will easily chip and peel, if not rubbed down or cut back.

- ◆ Knot and prime any eased edges or repairs. When dry, rub down to blend in primer to existing paint surface.
- ◆ Fill and rub down any minor defects and imperfections.
- ◆ Remove surface dust and apply undercoat followed by the top coat within recommended over-coating time.

When the old system has broken down, it is best to completely strip off the old paint, make good and start again from scratch using the same procedure as for bare wood.

Small areas showing signs of deterioration may be repaired without fully stripping the area.

- ◆ Treat any minor areas of soft timber caused by wet rot with a wood hardener.
- ◆ Rub down the surface to remove all loose defective paint.
- ◆ Apply the paint system as before.

Painting procedure

Internal painting

Protect carpets and furniture with dustsheets. Doors to other rooms can be sealed with masking tape prior to any rubbing down. Open the window to ensure adequate ventilation. This is both for you and to help the paint dry. Always wear a dust mask when rubbing down and eye protection goggles when scraping off.

- ◆ Primers and undercoats are applied by brushing out along the grain.
- ◆ Top coats are initially applied along the grain, brushed out across the grain and finally 'laid off' or finished with gentle brush strokes along the grain. The aim is to produce a thin, even paint film, which does not 'sag' on vertical surfaces or 'pond' on horizontal ones.

External painting

In general the same procedure for internal painting can be adopted, except for adverse weather conditions, and taking extra care to ensure full paint coverage to avoid the possibility of moisture penetration.

- ◆ Do not work in strong sunlight, as this prevents paints drying properly and makes it likely to cause it to 'blister'. Wait until the area is in shade before painting.
- ◆ Do not paint if rain is expected.
- ◆ Do not paint first thing in the morning or last thing in the evening, when there might be 'dew'. The resulting moisture will spoil the paint finish.
- ◆ Do not paint when there is a risk of frost.
- ◆ Do not use paint intended for interior use only.

In these circumstances it is best to aim to paint from mid-morning to just after lunch. This will allow the air to dry before starting and the paint film to dry before the early evening dampness starts to form.

Doors: remove handles, paint in sequence shown. Leading edge should be painted to match the woodwork of the room it opens into

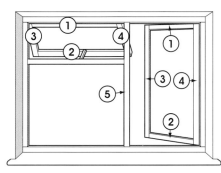

Casement window: paint opening parts before frame and interior sill

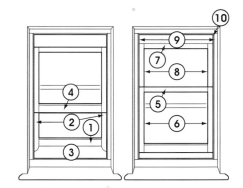

Sash window: from inside open sashes as far as they will go, paint all accessible surfaces, reverse sashes and complete painting

Figure 5.38
Sequence for painting doors and windows

Sequence of operations

A logical approach is required when painting framed joinery. The aim is to keep a 'wet edge' blending in adjacent areas of paint, so that joints are not seen when the paint dries.

Figure 5.38 illustrates typical numbered sequences for a range of joinery.

Painting plasterwork

Walls and ceilings are normally painted using emulsion paint. Newly plastered and repaired surfaces should be left for seven to 10 days to dry and then treated with a coat of plaster sealer before decoration. This prevents the new plaster showing through the paint finish as a kind of 'patchiness'. Alternatively, a thinned emulsion can be applied as a primer, before at least two full-strength coats are put on. The priming coat should be about one part water to about three parts emulsion paint.

Before starting work, arrange for any furniture in the room to be removed. Protect carpets and fixtures with dustsheets. Wear a dust mask and eye protection goggles when rubbing down and scraping off.

Maintenance Chapter 5

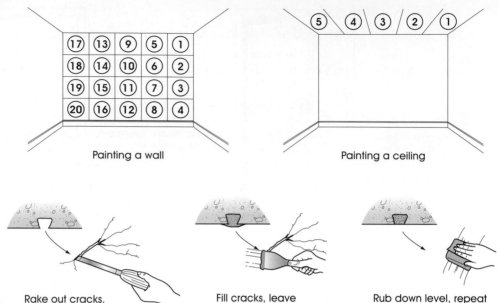

Figure 5.39
Sequence for painting and filling plasterwork

Painting a wall

Painting a ceiling

Rake out cracks, undercut edges

Fill cracks, leave filler proud of surface

Rub down level, repeat fill and rub down if required

The procedure to follow is illustrated in Figure 5.39.

◆ Remove all loose material such as dirt, dust and flaking paint.
◆ Rake out any minor cracks in the plaster surface, using the end of an old slot blade screwdriver. The raking out is to make the crack a little deeper and wider with undercut edges.
◆ Fill cracks with a plaster filler. Ensure filler is pressed well into the cracks in order for it to key on the undercut edges.
◆ Leave filler slightly proud of surrounding surfaces. Rub down level when dry. Fill the area again and rub down if required.
◆ Rub down the entire wall or ceiling surface.
◆ Wash down the surface and allow it to dry.
◆ Ceilings should be painted before walls. A small brush is used to cut into the corners and up to the frames and skirtings, etc.
◆ Use a roller or large brush to cover an area of about 1 m² at a time. Apply paint in one direction and spread it out by brushing or rollering diagonally. Finally finish off using light pressure only, in the same direction as you started.
◆ Using the numbered sequence continue painting the subsequent squares or strips, blending in the paint application of one with another while the paint is still wet. Otherwise pronounced lines will be apparent in the finished work, if wet paint is applied over a drying one.
◆ Clean all brushes/rollers and equipment in water on completion.

Paint coverage

This depends upon the absorbency of the surface to be painted and the quality of the paint. Typically:

◆ primers and undercoats cover 12–14 m² per litre per coat;
◆ gloss or satin top coats cover 14–16 m² per litre per coat;
◆ emulsions cover 10–12 m² per litre per coat.

Guttering and downpipes

As a carpenter and joiner you may be required to clean out fallen leaves and twigs from gutters or replace damaged or worn out guttering and down pipes while undertaking other maintenance activities.

Gutters are used to collect rainwater that falls on the roofs of buildings. It flows from the gutter into a down or rainwater pipe (RWP), which is connected either directly to a back inlet gulley, or over an open grated gulley. In situations where the rainwater pipe discharges into an open grated gulley the RWP will be fitted with a shoe to ease/direct the flow of rainwater as illustrated in Figure 5.40.

Most gutters and rainwater pipe systems installed in new domestic properties and for replacement tend to be plastic, whereas in the past cast iron and asbestos cement was used. You may even come across solid timber section gutters known as spouting.

If when working on a maintenance job you suspect that the gutters or downpipes are made of asbestos cement you should not attempt any repair and your supervisor should be called immediately for advice as to the correct procedures to follow.

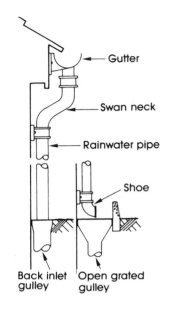

Figure 5.40 *Rain water pipe to gulley connections*

Gutters

Gutters are specified by the shape of their cross-section as illustrated in Figure 5.41. Also shown in the figure is a range of fittings. Gutters are available in 2 m and 4 m lengths and range from 75 mm to 150 mm in width. Conversion fittings are normally available to connect new plastic lengths of guttering with existing cast iron sections. Gutters are normally laid at a fall of 1:600; that is 1 mm out of level for every 600 mm run.

Rainwater pipes

Rainwater pipes are either square or round in section. They are available in lengths of 2.5 m, 4 m and 5.5 m. For most domestic replacements either 65 mm square or 68 mm diameter are used.

Asbestos is particularly harmful to your health, especially if its dust or fibres are inhaled. You may come across asbestos when working on maintenance jobs. Before undertaking major refurbishment or maintenance jobs, contractors should have an asbestos survey undertaken. The removal of asbestos or working in an area of asbestos must only be carried out by specialist companies.

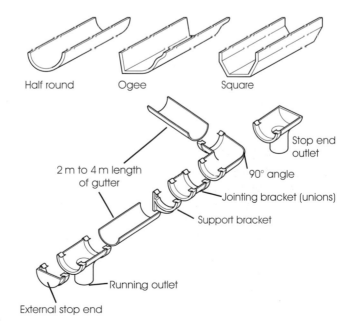

Half round   Ogee   Square

2 m to 4 m length of gutter

Stop end outlet

90° angle

Jointing bracket (unions)

Support bracket

Running outlet

External stop end

Figure 5.41 *Gutter profiles and associated fittings*

Maintenance   Chapter 5

safety tip

A proper working platform is required when replacing gutters; working off a ladder is not considered appropriate and may contravene the Working at Height Regulations. Your employers should undertake a risk assessment before anyone starts to work at heights.

Replacing gutters and rainwater pipes

The following general procedure can be used when replacing guttering (see Figure 5.42):

- Old cast-iron gutters are very heavy and difficult to manage and in addition the bolts used to join sections will probably have rusted up. You will have to cut off the bolts with a hacksaw, and remove and lower the gutter to the ground length by length, before any replacement can start.
- Use a plumb line centred over the gulley to position the outlet.
- Fix outlet to the fascia board; the gap between the top edge of the gutter and the bottom of the roof tile should be a maximum of 30 mm and in addition any tile underfelt should extend into the gutter.
- Work out the fall over the length of the gutter and fix the last bracket. This amount above the outlet, e.g. a 6 m run of guttering, will require a fall of at least 10 mm.
- Fix a string line between outlet and end bracket; use the line to position either plain brackets or union brackets at a maximum of 1 m centres.
- Once support brackets are in place, lengths may be cut to suit and snapped into position. Make sure when cutting that you allow for thermal movement about 3 mm per metre run. Most plastic unions have an 'insert to here' mark on the inside. Cutting too log will result in expansion, buckling and strange noises in hot weather. Cutting too short may result in the joint coming apart or leaking in cold weather due to contraction.
- The final length of gutter will require a stop end; this should be positioned about 50 mm beyond the roof tiling, to aid the collection of wind-driven rain.
- A sawn neck or offset is normally required at the upper up of the rainwater pipe to clear the width of the eaves soffit and allow the pipe to be clipped back to the surface of the wall. These are in two pieces that require cutting to the appropriate length. About 6 mm should be allowed for expansion.
- Once the swan neck is in place lengths of downpipe can be positioned and clipped in place with pipe brackets at around 1 m intervals. Check with a spirit level for plumb before drilling brackets. Again about 6 mm should be allowed between the end of one pipe and the inside shoulder of the fitting for expansion.
- Finally either connect the downpipe to the gully or fit a shoe.

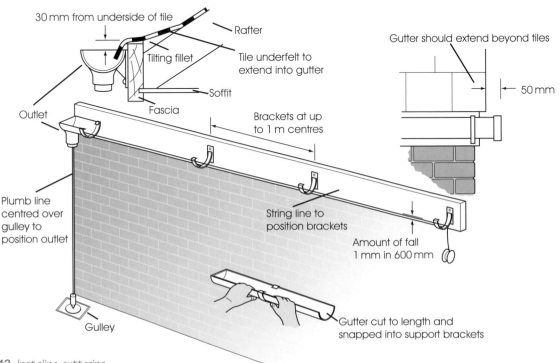

Figure 5.42 *Installing guttering*

activity

(Reference to *A Building Craft Foundation* (3rd edition) may be required to complete this task.)

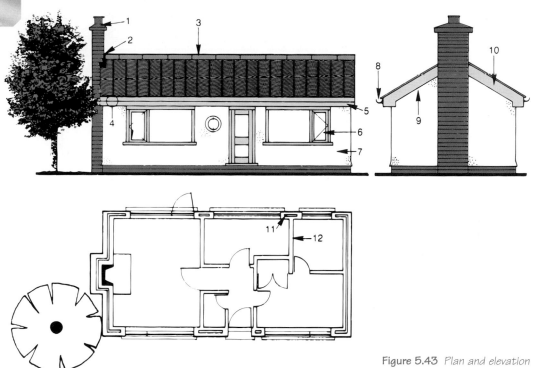

Figure 5.43 *Plan and elevation*

1. Name the type of accommodation illustrated in Figure 5.43.

2. Using the following items, identify the numbered features shown in Figure 5.43; not all of the items are applicable: gable, ridge, sash, hip, verge, eaves, parapet, cavity, partition wall, cladding, lintel, flashing, casement, bargeboard, fascia board, gutter, rendering, casement, flaunching.

3. Define the following terms and indicate an example of each of them on the section shown in Figure 5.44:

   (a) substructure
   (b) superstructure
   (c) primary element
   (d) secondary element
   (e) finishing element
   (f) component.

4. Name the type of foundations illustrated in Figure 5.44.

5. Sketch an alternative type of ground-floor construction.

6. Identify the lettered elements/components shown in the section in Figure 5.44 and name the material indicated.

Maintenance **Chapter 5**

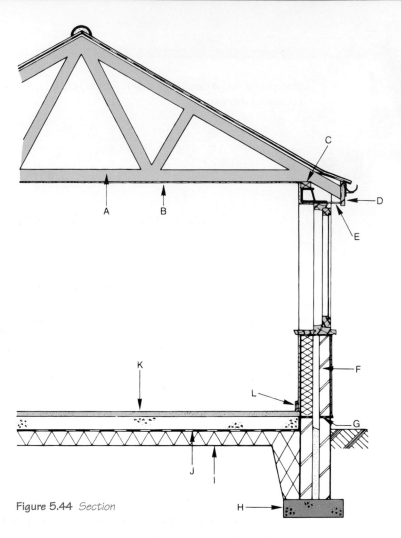

Figure 5.44 *Section*

7. During a close inspection of the building in Figure 5.44 you notice the following defects. State a possible cause and remedy for each:

    a) Small damp patch in the centre of the lounge floor
    b) Rafters next to chimney stack are wet and show signs of fungal attack
    c) Vertical cracks both internally and externally in walls down the side of the chimney and under the lounge window
    d) Soft woodwork to external kitchen window sill
    e) Front door has dropped at head, sticks on the threshold and shows signs of open/loose joints.

8 Write a letter to the building owners, informing them of the defects you identified during your survey visit and suggest appropriate remedial action.

1. Name the work activities associated with the following operations:
   (a) Pointing
   (b) Splicing
   (c) Grouting
   (d) Ruling off
   (e) Cutting in
   (f) Laying off
   (g) Trowelling up.

2. Name THREE agents of deterioration in buildings.

3. State why screwing of plasterboard when repairing a ceiling is preferable to nailing.

4. A high level sash window requires re-glazing. Describe a safe method of work.

5. You are asked to replace a badly decayed window frame with a new one. However, on examination there is no lintel or other means of support evident above the opening. Outline the procedure to follow.

6. On lifting a floorboard, the underside is found to be soft, powdery and full of small holes. The most likely cause is:
   (a) wet rot attack
   (b) dry rot attack
   (c) wood-boring insects
   (d) excessive floor load.

7. A maintenance carpenter has a 'mouse' in his tool bag. State what it would be used for.

8. State or sketch the sequence of operations required to keep a 'wet edge' when applying the finishing coat of paint to a six-panel door.

9. State the precautions to be taken before removing unsound plasterwork in a carpeted room.

10. Describe the procedure for disposing of old timber, shavings and swept up dust during the eradication of a dry rot attack.

11. Produce a sketch to show how an old cylinder lock recess and hole in a door can be made good when changing lock positions.

12. Explain the difference between surface and interstitial condensation.

13. Define the difference between planned maintenance and corrective maintenance.

14. State the procedure to be followed if you suspect that the guttering you have to replace is made from asbestos.

15. State the normal amount of fall that guttering should have.

6

Circular Saws

This chapter is intended to provide the reader with an overview of setting up and using circular saws in both a workshop and on-site situation. Its contents are assessed in the **NVQ Unit VR 13 Set Up and Use Circular Saws.**

In this chapter you will cover the following range of topics:

- legislation and guidance;
- accident statistics;
- types of circular saw;
- safe working practices for circular saws;
- tooling;
- maintenance;
- troubleshooting.

What is required in VR 13 Set Up and Use Circular Saws?

To successfully complete this unit you will be required to demonstrate your skill and knowledge of the following processes:

- Interpreting information;
- Adopting safe and healthy working practices;
- Selecting materials, components and equipment;
- Setting up, using and maintaining circular saws in accordance with current legislation and official guidance.

You will be required practically to:

- use fixed or transportable circular saws and appropriate aids to:
  - ▶ cut timber and timber manufactured sheet material
  - ▶ change saw blades;
- use appropriate personal protective equipment to carry out an activity;
- maintain a clean work area and safely dispose of waste material;
- identify problems associated with the use of circular saws;
- communicate with other team members;
- undertake calculations for quantity, length, area and wastage.

Legislation and guidance

Practical guidance on safe working practices to be observed when using woodworking machines is contained in an Approved Code of Practice (ACOP) *Safe Use of Woodworking Machinery.* This ACOP takes into account both the practical aspects of machine safety and the legal requirements contained in the Provision and Use of Work Equipment Regulations (PUWER) and the Management of Health and Safety at Work Regulations (MHSW). In addition, the Health and Safety Executive produce a series of *wood information sheets*, which contain further practical guidance on the safe use of individual woodworking machines. These may be viewed on this website: www.hse.gov.uk.

Duties in legislation can be either be **absolute** or have a qualifying term added called **reasonably practicable**.

▶ **Absolute** – unless a qualifying term is added such as **whenever reasonably practicable**, the requirement must be met regardless of cost or any other consideration.

▶ **Reasonably practicable** – means that you are required to consider the risks involved in undertaking a particular work activity. However, if the risks are minimal and the cost or technical difficulties of taking certain actions to eliminate the risks are very high, it might not be reasonably practicable to take those actions.

The main general requirements to be considered wherever woodworking machines are to be used may be summarised under the following headings:

Use of safety appliances

◆ The use of safety appliances, such as push sticks and jigs, keep the operator's hands in a safe positions, whilst allowing the operator to maintain full control of the workpiece during cutting operations.
◆ Power feeds reduce the need for hands to approach the cutters and should be used whenever reasonably practicable.

Machine controls

◆ All machines should be fitted with a means of isolation from the electric supply, which should be located close to the machine.
◆ Lockable isolators can be used to prevent unauthorised use of a machine and give increased protection during maintenance.

Working space

◆ Machines should be located in such a position that the operator cannot be pushed, bumped or easily distracted.
◆ There should be sufficient space around machines for the items to be machined, for finished workpieces and for waste bins, so that there is no obstruction affecting the operator.
◆ Wherever possible, machine shop areas should be separated from assembly or packaging areas, and from areas used by forklift trucks or other forms of transportation.
◆ All access and escape routes must be kept clear.
◆ Waste bins should be emptied at regular intervals and waste sacks containing wood dust should be stored outside the workroom.

Floors

◆ The floor surface of the work area must be level, non-slip and maintained in good condition.
◆ The working area around a machine must be kept free from obstruction, off-cuts and shavings, etc.
◆ Supply cables and pipes should be routed at high level or set below floor level, in order to prevent a tripping or trapping hazard.
◆ Polished floors should be avoided as they present a risk of slipping.
◆ All spillages should be promptly cleared up to avoid the risk of slipping.

Lighting

◆ Machinists require good lighting (natural or artificial) in order to operate safely.
◆ Lighting should be positioned or shaded to prevent glare and not shine in the operator's eyes.
◆ Adequate lighting must also be provided for gangways and passages.
◆ Windows should be shaded when necessary to avoid reflections from worktables and other shiny surfaces.

Heating

◆ Low temperatures can result in a loss of concentration and cold hands can reduce the operator's ability to control the workpiece safely.
◆ A temperature of 16°C is suitable for a machine shop.
◆ In a sawmill where heavier work is undertaken a temperature between 10°C and 16°C is considered suitable.
◆ Where it is not possible to heat the entire area, radiant heaters can be provided near or adjacent to the work area to enable operators to warm themselves periodically.

Dust collection

◆ Wood dust is harmful to health.
◆ Woodworking machines should be fitted with an efficient means of collecting wood dust and chippings.
◆ Local exhaust ventilation systems should be regularly maintained to prevent their efficiency from deteriorating.

Training

◆ Individuals should not use any woodworking machine unless they have been properly trained for the work being carried out.
◆ People under 18 years of age are prohibited from operating certain machines, unless they have successfully completed an approved training course.

Accident statistics

When compared to other industries woodworking accounts for a disproportionately high number of machine accidents. In a Health and Safety Executive (HSE) survey of 1000 woodworking machine accidents circular saw benches accounted for 35% of the total as illustrated by the pie chart in Figure 6.1.

Many of these accidents resulted in the loss of fingers. Of the total for circular saw benches, 83% occurred whilst undertaking ripping or crosscutting operations. In most cases the saw guard was either incorrectly adjusted or missing altogether.

The HSE states that many of the accidents could have been avoided if the saw guard was correctly adjusted and a push stick used.

The HSE statistics go on to show that accidents are disproportionately high in premises that employ a smaller number of people as illustrated in Figure 6.2. Over 50% of those injured had only received 'on-the-job training'; 24% had not received training or instruction on the machine they were using and of these only 5% were under supervision; finally 25% of accidents involved formally trained operators, which indicates that safe working methods were being bypassed.

Risks must be controlled by:

▶ **eliminating the risk** – or if that is not possible:

▶ **taking hardware measures** such as the **provision of guards** to control the risks; but if the risks cannot be adequately controlled:

▶ **taking software measures** such as **following safe systems of work** and the provision of information, instruction and training to deal with the remaining risk.

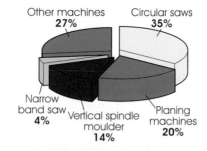

Figure 6.1 *HSE survey of woodworking machine accidents*

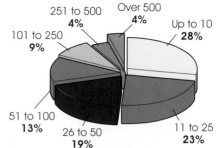

Figure 6.2 *HSE statistics showing percentage of accidents by size of firm*

To help prevent accidents at woodworking machines all concerned should:

◆ Assess all the risks in the workplace and put precautions in place to eliminate them.
◆ Ensure all necessary guards are in position and used at all times.
◆ Ensure all machine operators are suitably trained on the machine they are using and, in addition, ensure that they are properly supervised.
◆ Check that machine operators are following safe working methods at all times.

Types of circular saw

The three main types of circular saw are:

1. **Cross-cut saw** – used for cutting to length.
2. **Rip saw** – used for cutting to width and thickness.
3. **Dimension saw** – used for the precision cutting of timber and sheet material.

There are two basic types of sawing operation as illustrated in Figure 6.3. In the first, used for cross cutting only, the material, which is being cut remains stationary on the table while the revolving saw is drawn across it. The second type, where the material is fed past the revolving saw, is suitable for both rip and cross cutting.

Figure 6.3 *Circular saw (basic methods)*

Cross-cut saw

The saw unit is drawn across the material, cutting it to length (Figure 6.4).

Adjustable length stops may be fitted where repetitive cutting to the same length is required.

With most models it is also possible to carry out the following operations (see Figure 6.5):

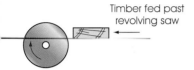

◆ cross cutting;
◆ compound cutting;
◆ cutting birdsmouths;
◆ cutting housings;
◆ cutting notches;
◆ cutting halving joints;
◆ kerfing;
◆ ripping (with riving knife fitted);
◆ trenching, tenoning and ploughing with special cutters.

Figure 6.4 *Cross-cut saw*

Circular Saws | Chapter 6

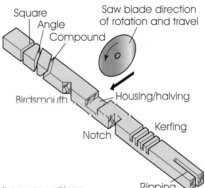

Figure 6.5 *Cross-cut saw operations*

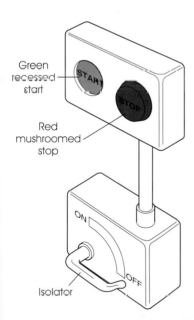

Figure 6.6 *Start, stop and isolation controls*

Starting, stopping and isolation controls

In common with most woodworking machines, a recessed start button and a mushroom-head stop button control the motor. When connected up, all machines should also be fitted with an isolating switch, so that the machine can be completely isolated (disconnected) from the power supply when setting, adjusting, or carrying out maintenance work on the machine (see Figure 6.6). The purpose of recessing the start button is to prevent accidental switching on. The stop button is mushroomed to aid positive switching off and should be suitably located to enable the operator to switch the machine off with a knee in an emergency.

In addition, to reduce the risk of contact with the operator during the rundown, machines should be fitted with a braking device that brings blades and cutters safely to rest within 10 seconds.

Rip saw

The material is fed past the revolving blade on the rip saw to cut it to the required section (see Figure 6.7).

Two main operations are involved in cutting timber to the required section (as illustrated in Figure 6.8):

Figure 6.7 *Rip saw*

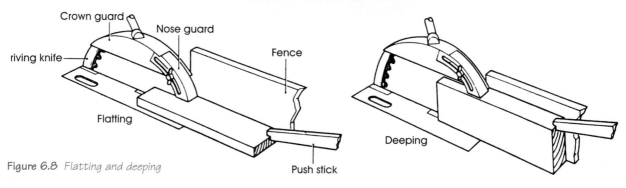

Figure 6.8 *Flatting and deeping*

1. Cutting the timber to the required width, which is known as flatting.
2. Cutting the timber to the required thickness, which is known as deeping.

In addition to flatting and deeping, a third operation may be required as illustrated in Figure 6.9.

3. Cutting the timber to the appropriate bevel, taper or wedge shape. Machine operators often make up their own saddles, bed pieces and jigs to enable them to safely carry out bevel, angle and taper cutting.

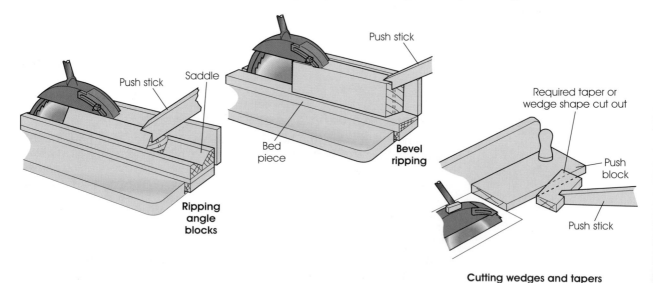

Figure 6.9 *Bed pieces, saddles and jigs*

Some saws have a recess on each side and in front of the blade where it enters the table. These recesses are intended to receive felt packings and a hardwood mouthpiece. The packing helps to keep the saw cutting in a true line. The mouthpiece helps to prevent the underside of the timber breaking out or 'spelching'.

When setting the machine up for any ripping operation, the fence should be adjusted so that the arc on its end is in line with the gullets of the saw teeth at table level. Binding will occur if it is too far forward and, if it is too far back, the material will jump at the end of the cut, leaving a small projection.

Dimension saw

did you know?

Dimension saws are also referred to as 'panel saws'.

This is used for cutting timber and sheet material to precise dimensions (see Figure 6.10). Most sawing operations are possible although on a lighter scale than the previous two machines.

Figure 6.10 *Dimension saw*

The cross-cut fence, when fitted, adjusts for angles and the blade may be tilted for bevels/compound cutting and can be moved up and down. The large sliding side table, used for cross cutting, also serves to give support when cutting sheet material.

Safe working practices for circular saws

Cross-cut saws

The guidance applicable to cross-cut saws is illustrated in Figure 6.11 and summarised in the following points:

1. The non-cutting part of the blade must be totally enclosed with a fixed guard, which should extend down to at least the spindle.
2. Guards or a saw housing should be provided so that there is no access to the saw blade when in its rest position.
3. A nose guard should be fitted to prevent contact with the front edge of the blade during cutting and when the saw is at rest.
4. The maximum extension or stroke of the saw should be set so the nose guard cannot extend beyond the front of the saw table.
5. A braking device should be fitted to the machine that brings the blade to rest within 10 seconds, unless there is no risk of contact with the blade during rundown.
6. A fence is required on either side of the cutting line and should be high enough to support the timber being cut. The gap in the fence should be just sufficient to allow the passage of the nose guard. When straight cutting

on a machine that is capable of angled cuts, any excessive gap in the fence should be closed by the use of renewable fence inserts or false fence.

7. It is recommended that 'no hands' areas be marked in yellow hatching on the table 300 mm either side of the blade. Operators should be trained not to hold timber in these areas during cutting operations.

8. Workpiece holders or jigs should be used when cutting small workpieces or narrow sections.

9. Offcuts and woodchips should only be removed when the saw has stopped and is in the rest position; even then it is good practice to use a pushstick rather than the hands.

10. In order to reduce the likelihood of distorted timber binding on the saw causing kickback, any bow should be placed against the bed and any spring against the fence, with packers being used to prevent rocking.

11. Although some machines have the facility to turn the cutting head through 90° to allow rip sawing, a circular saw bench is considered a safer, more suitable option.

12. Jigs and workpiece cramps should be used when undertaking operations such as trenching and the pointing of stakes or pales in order to provide workpiece stability and prevent kickback.

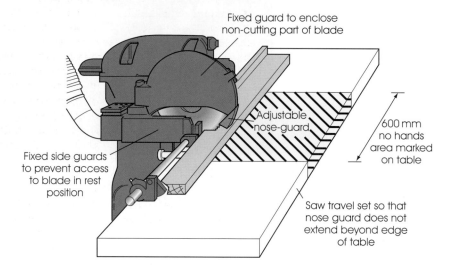

Elevation

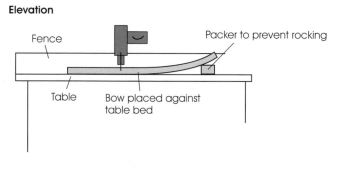

Plan

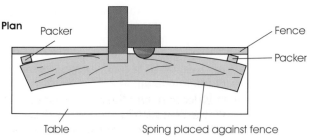

Figure 6.11 *Safe use of manually operated cross-cut saws*

Circular saw benches

The guidance applicable to circular-saw benches is illustrated in Figure 6.13 (see page 222) and summarised in the following points:

1. The part of the saw below the saw table must be fully enclosed.
2. In order to reduce the risk of contact with the moving saw blade during rundown a braking device must be fitted to the machine that brings the blade to rest within 10 seconds.
3. A riving knife must be fitted directly behind the saw blade. Its purpose is to part the timber as it proceeds through the saw and thus prevent it jamming on the blade and being thrown back towards the operator. Whenever the saw blade is changed the riving knife must be adjusted so that it is as close as practically possible to the saw blade and, in any case, the distance between the riving knife and the teeth of the saw blade should not exceed 8 mm at table level. The distance between the top of the riving knife and the top of the blade should be no more than 25 mm, except for blades over 600 mm diameter where the riving knife should extend at least 225 mm above the table. Riving knifes should have a chamfered leading edge and be thicker than the saw blade but slightly thinner than the width (kerf) of the saw cut.
4. The upper part of the saw blade must be fitted with a strong adjustable saw guard (crown guard), which has flanges on either side that cover as much of the blade as possible and must be adjusted as close as possible to the workpiece during use. An extension piece (nose guard) may be fitted to the leading end of the crown guard. This guards the blade between the workpiece and crown guard. If when cutting narrow workpieces the guard cannot be lowered sufficiently because it fouls a fixed fence, a false fence should be fitted. In all circumstances the extension piece must be adjusted as close as possible to the workpiece during use.
5. The diameter of the smallest saw blade that can safely be used should be marked on the machine. A smaller blade, less than 60% of the largest saw blade for which the machine is designed, will not cut efficiently due to its lower peripheral (tip of teeth) speed.
6. Saw benches should be fitted with local exhaust ventilation above and below the table, which effectively controls wood dust during the machine's operation.
7. Where an assistant is employed at the outfeed (delivery) end of the machine to remove the cut pieces, an extension table must be fitted so that the distance between the saw blade spindle and the end of the table is at least 1200 mm. The assistant should be instructed to remain at the outfeed end of the extension table and not to reach forward towards the saw blade.
8. The operator's hands should never be in line with the saw blade or be closer than necessary to the front of the saw. A suitable push stick (Figure 6.12) should be used in the following circumstances:
   ► feeding material where the cut is 300 mm or less;
   ► feeding material over the last 300 mm of the cut;
   ► removing cut pieces from between the saw blade and fence unless the width of the cut piece exceeds 150 mm.

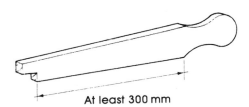

At least 300 mm

Figure 6.12 Push stick

9. In order to reduce the risk of contact with the saw blade, it is recommended that a demountable power feed is used whenever possible. This is not a substitute for the riving knife, which must be kept in position at all times.

Circular Saws **Chapter 6**

10. A fence should always be used to give support to the workpiece during cutting. For shallow or angled cutting the normal fence may need replacing with a low fence to enable the use of a push stick or prevent the canted blade touching the fence.

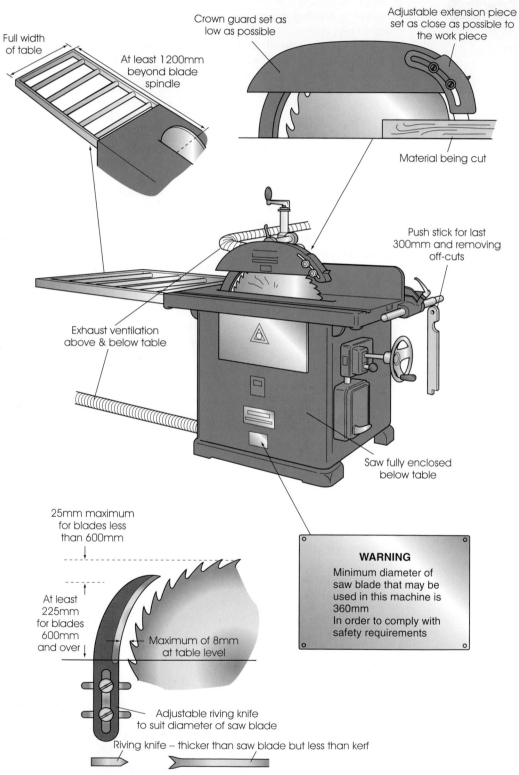

Full width of table

At least 1200mm beyond blade spindle

Crown guard set as low as possible

Adjustable extension piece set as close as possible to the work piece

Material being cut

Push stick for last 300mm and removing off-cuts

Exhaust ventilation above & below table

Saw fully enclosed below table

25mm maximum for blades less than 600mm

At least 225mm for blades 600mm and over

Maximum of 8mm at table level

WARNING
Minimum diameter of saw blade that may be used in this machine is 360mm
In order to comply with safety requirements

Adjustable riving knife to suit diameter of saw blade

Riving knife – thicker than saw blade but less than kerf

Figure 6.13 *Circular saw safety requirements*

11. The safe working position for the operator is at the feed end offset away from the fence and out of the blade line. See Figure 6.14.

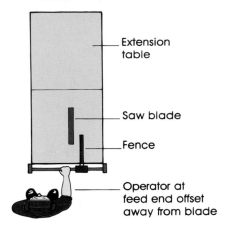

- Extension table
- Saw blade
- Fence
- Operator at feed end offset away from blade

Figure 6.14 Saw operator position

12. It is recommended that operators wear personal protection (see Figure 6.15): ear protection to reduce the risk of hearing loss; dusk mask or respirator, particularly when cutting hardwoods to reduce the risk of respiratory problems; and goggles or a face screen where there is a risk of flying particles.

Ear muffs   Dust Mask   Respirator   Goggles   Face screen

Figure 6.15 Use personal protection when machining

13. Circular saws must not be use for the following operations:
- ◆ Cutting tenons, grooves, rebates or mouldings unless effectively guarded. Guards normally take the form of Shaw 'tunnel-type' guards, which, in addition to enclosing the blade, apply pressure to the workpiece, keeping it in place.
- ◆ Ripping is not permissible unless the saw teeth project above the timber, e.g. deeping large sectioned material in two cuts is not allowed.
- ◆ Cutting round section timber unless the workpiece is adequately supported and held by a gripping or cramping device.
- ◆ Angle cuts and bevels can be made on tilting arbour saws by inclining the blade; the fence should be set in the low position or a false fence used to prevent the rotating blade coming into contact with it. On non-tilting fixed position saws bed pieces or jigs can be used to provide workpiece support during cutting.

Circular Saws · Chapter 6

Tooling

did you know?

Employers have a duty to reduce the hazards created by dust, noise and vibration in the workplace.

▶ *Dust.* Carry out assessment of health risks and prevent exposure to dust or, where this is not reasonably practicable, take adequate measures to control it, including local dust extraction and the use of PPE.

▶ *Noise.* Carry out assessment of levels and those at risk. Designate noisy areas as ear protection zones. Provide ear protection and ensure it is worn by all who work in or enter the protection zone.

▶ *Vibration.* Undertake health surveillance of those exposed; take preventative measures to reduce exposure including limiting the duration of exposure, use of reduced vibration machinery, adaptation of safer working methods and the use of anti-vibration gloves.

A range of **circular saw blades** is available for various types of work. Figure 6.16 illustrates the section of two in common use:

◆ **plate,** also known as **parallel plate,** for straightforward rip and cross cutting work;
◆ **hollow ground** for dimension sawing and fine finished work.

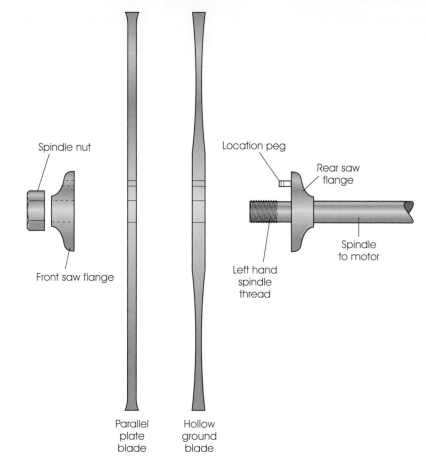

Figure 6.16 *Circular-saw blade section*

Other section blades with a thin rim are available but have limited uses, for example, swage, ground-off and taper. They are used for rip sawing thin sections. Each has its own particular application, although the purpose of each is the same, i.e. to save timber by reducing the width of the saw kerf.

Saw teeth

Saw teeth require setting so that the kerf (width of the saw cut) produced is wider than the thickness of the blade. Otherwise it will bind on the timber and overheat as a result of the friction, causing the blade to wobble and produce a wavy or 'drunken' cut.

The teeth can be set in two main ways, as shown in Figure 6.17:

◆ *Spring set teeth – where adjacent teeth are sprung to the opposite side of the blade. This is the same method as that used for handsaws.*
◆ *Swage set teeth – mainly used for setting thin rim rip saws. The point of each tooth is spread out evenly on both sides to give it a dovetail-shaped look.*

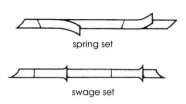

Figure 6.17 *Types of saw blade setting*

did you know?

'TCT' is used as an abbreviation for tungsten carbide tipped saw blades.

Hollow ground and tungsten tipped saws do not require setting as the hollow grinding provides the necessary clearance or the tip side overhang respectively.

Teeth shape

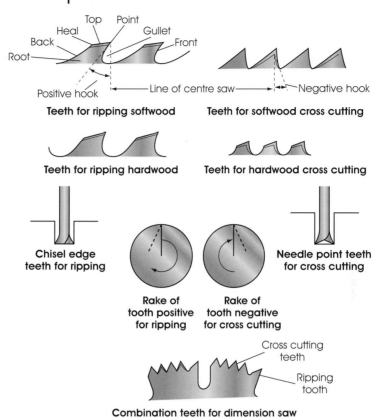

Figure 6.18 *Circular saw teeth*

For efficient cutting the shape of the saw teeth must be suitable for the work being carried out (see Figure 6.18).

◆ **Rip saws** require **chisel-edge teeth**, which incline towards the wood (they have positive hook). Teeth for ripping hardwood require less hook than those for ripping softwood.
◆ **Cross-cut saws** require **needle-point teeth**, which incline away from the wood (they have negative hook). The needle-point teeth for hardwood cross-cutting must be strongly backed up.
◆ **Dimension sawing** ideally requires a **combination blade** of both rip and cross-cut teeth, although, as dimension saw benches are rarely used for ripping, a fine cross-cut blade is often fitted.

Tungsten carbide tips

The use of wear-resistant tungsten carbide tipped teeth saws (Figure 6.19) is recommended when cutting abrasive hardwoods, plywood, MDF and chipboard. Their use reduces excessive blunting of the saw, thus extending the period before blade changing and resharpening is required.

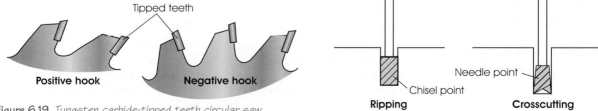

Figure 6.19 *Tungsten carbide-tipped teeth circular saw*

Maintenance

did you know?

The routine periodic maintenance of a machine is also termed 'preventative maintenance' as, during this work, worn or damaged parts can be identified and replaced before they break down.

Saw blade maintenance

After a period of use, saw blades will start to dull (lose their cutting edge). This will progressively cause a poor finish to the saw cut, including burning of both the timber and the blade and possibly cause blade wobble due to overheating. In addition, it will require excessive pressure by the operator to force the timber through the saw.

The sharpening of circular blades is normally carried out on a saw sharpening machine or by hand filing. However, neither of these operations is within the scope of this Unit of Competence.

To ensure true running of a saw blade, it should be fitted in the same position on the saw spindle each time it is used. This can be achieved by always mounting the blades on the spindle with the location/driving peg uppermost and, before tightening, pulling the saw blade back onto the peg.

Resin deposits on saw blades should be cleaned off periodically. They can be softened by brushing with an oil/paraffin mixture and scraped off. A wood scraper is preferable, as it will avoid scratching the saw blade.

Machine maintenance

Routine periodic maintenance of the machine will:

◆ prolong its serviceable life;
◆ ensure all moving parts work freely;
◆ ensure the machine operates safely.

The manufacturer's maintenance schedule, supplied with each machine, gives the operator information regarding routine maintenance procedures. The schedule will detail the parts to be lubricated, the location of grease nipples and the type, the required frequency and amount of grease.

A typical procedure might be:

1. Remove all rust spots with fine wire wool.
2. Clean off resin deposits and other dirt, using an oil/paraffin mixture and wooden scraper.
3. Wipe over entire machine using clean rag.
4. Apply a coat of light grade oil to all screws and slides. Excess should be wiped off using a clean rag.
5. Clean off grease nipples and apply correct grade and amount of grease using the correct gun. Parts can be rotated manually during this operation.
6. Check freeness of all moving parts.

Troubleshooting

The most common faults that occur when sawing timber along with their probable causes and suggested remedies are listed in Table 6.1.

Table 6.1 Circular sawing: faults, causes, remedies

Fault	Probable cause	Remedy
The saw blade begins to wobble	Saw blade overheating due to: ◆ packing too tight ◆ dull teeth ◆ insufficient set ◆ abrasive timber ◆ loss of tension in saw blade	◆ Reduce thickness of packing ◆ Replace with sharp saw blade ◆ Set teeth ◆ Use tungsten-tipped saw blade ◆ Replace saw blade
The timber being sawn moves away from the fence or binds against the fence	◆ Fence not parallel to saw blade ◆ Arc on fence not set in line with gullets of teeth	◆ Realign fence ◆ Adjust arc of fence to line up with gullets of teeth
Rough sawn finish	◆ Uneven setting or sharpening of teeth	◆ Replace with correctly sharpened and set saw blade
The blade binds in the saw kerf	◆ Dull teeth ◆ Insufficient set ◆ Case hardened or twisted timber	◆ Replace with sharp saw blade ◆ Replace with correctly set saw blade ◆ Avoid if possible; if not, use tungsten-tipped blade and feed slowly forward, easing back when binding occurs
Small projection left at the end of timber after ripping	◆ Arc on fence set too far back from line of gullets, causing the material to jump into gap at end of operation	◆ Realign arc on fence with the saw teeth gullets at table level
Saw cut not square to either table when ripping or fence when crosscutting	◆ Saw blade set at an angle to the table ◆ Cutting head set at an angle to the fence	◆ Adjust inclination of saw blade so that it is at right angles to the table ◆ Adjust cutting head so that it is at right angles to the fence

Circular Saws **Chapter 6**

Consult The Provision and Use of Work Equipment Regulations (PUWER) and the supporting Approved Code of Practice (ACOP) Safe Use of Woodworking Machinery, in order to answer the following questions:

1 List the **three** main duties that the regulations place on employers and the self employed who provide equipment for use at work or persons who control or supervise the use of equipment.
2 Explain what the regulations mean by **hardware and software** measures in relation to the control of risk to peoples health and safety created by the equipment that they use.
3 Describe what the ACOP says about young people and their use of woodworking machinery.
4 Describe what the ACOP says about training. Include in your answer who should be trained and three essential elements of a training scheme.
5 Explain the ACOP requirements under the following headings:
    (a) kickback   (b) disintegration
    (c) stop controls   (d) stability
    (e) markings   (f) warnings.

measuring up

1. Produce sketches to show the difference between deeping and flatting.

2. Name a type of saw blade that is most suitable for ripping abrasive timber.

3. Describe the safe working position that the operator of a circular hand fed saw bench should take.

4. The riving knife fitted to a circular saw must have a maximum clearance between itself and the blade at table level of:
   (a) 6 mm
   (b) 8 mm
   (c) 10 mm
   (d) 12 mm

5. The guard on a circular saw that covers the top of a saw blade is known as the:
   (a) shaw guard
   (b) top guard
   (c) crown guard
   (d) bridge guard

6. List **FOUR** general requirements for the safe use of woodworking machines.

7. State **ONE** piece of information that must be fixed to every circular, rip or dimension saw bench.

8. State the purpose of packings to circular saw blades.

9. State why a hardwood mouthpiece may be incorporated in the table of a circular saw bench.

10. Produce a labelled sketch of a circular saw blade tooth indicating six features.

11. State **TWO** reasons for undertaking routine periodic maintenance of woodworking machines.

12 State **TWO** situations where a push stick must be used.

13 List **FIVE** tasks that may be included in the periodic maintenance of a circular saw.

14 Explain why a riving knife thicker than the saw blade should be used.

15 Describe how you would ensure that a saw blade is refitted in exactly the same position after each time it has been taken off for sharpening.

Index